DAVID PROFUMO studied English at Magdalen College, Oxford, and King's College, London. His work has appeared in many newspapers and magazines, from the *Wall Street Journal* to *The Newsletter of the Kiribati* and *Tuvalu Philatelic Society*. The author of two novels, he was awarded the Geoffrey Faber Memorial Prize in 1989 and was elected a Fellow of The Royal Society of Literature in 1995. He has been Fishing Correspondent for *The Daily Telegraph* and *Country Life*, and lives in Perthshire.

Praise for The *Lightning Thread*

'An evocative, wide-ranging collection of essays from one of our finest writers, destined to be an angling classic . . . a remarkable book.' Neil Patterson, *Flyfishers' Journal*

'Unimaginable that any fisherman could put it down.' Thomas McGuane, author of *The Longest Silence*

'A fabulous confection of history, biology, philosophy and memoir . . . spiked with wit and crafted with precision and style.' Loyd Grossman

'An angling master, who wears his knowledge lightly . . . Everyone remotely interested in fishing, or writing, would love this book.' Prue Leith

'A writerly love of words is one thing that elevates *The Lightning Thread* above the status of mere memoir. Profumo relishes the craft of prose, shaping his sentences like casts which are often drawn deep from remote lexical pools.' Philip Marsden, *Spectator*

'Language is one of the loves of Profumo's life and he glories in its surprises and possibilities.' *Literary Review*

'Wonderfully well written . . . an absolute delight, and every page a satisfying surprise.' Sandy Leventon, *Trout & Salmon*

'With wit, quiet craft and a lifetime's
Profumo draws us into his paradis
the *Killing Eve* novels

Also by David Profumo

FICTION

Sea Music

The Weather in Iceland

NON-FICTION

The Magic Wheel (ed. with Graham Swift)

In Praise of Trout

Bringing the House Down: A Family Memoir

THE
LIGHTNING
THREAD

Fishological Moments and the
Pursuit of Paradise

DAVID PROFUMO

SCRIBNER
LONDON NEW YORK SYDNEY TORONTO NEW DELHI

First published in Great Britain by Scribner, an imprint of
Simon & Schuster UK Ltd, 2021

This edition published in Great Britain by Scribner,
an imprint of Simon & Schuster UK Ltd, 2022

Copyright © David Profumo, 2021

The right of David Profumo to be identified as the author of
this work has been asserted in accordance with the
Copyright, Designs and Patents Act, 1988.

SCRIBNER and design are registered trademarks of The Gale Group, Inc.,
used under licence by Simon & Schuster Inc.

1 3 5 7 9 10 8 6 4 2

Simon & Schuster UK Ltd
1st Floor
222 Gray's Inn Road
London WC1X 8HB

www.simonandschuster.co.uk
www.simonandschuster.com.au
www.simonandschuster.co.in

Simon & Schuster Australia, Sydney
Simon & Schuster India, New Delhi

The author and publishers have made all reasonable efforts
to contact copyright-holders for permission, and apologise
for any omissions or errors in the form of credits given.
Corrections may be made to future printings.

A CIP catalogue record for this book
is available from the British Library

Paperback ISBN: 978-1-4711-8657-8
eBook ISBN: 978-1-4711-8656-1

Typeset in Palatino by M Rules

Printed and bound by CPI Group (UK) Ltd, Croydon, CR0 4YY

For Helen, our lovely children James, Tom and Laura, and grandchildren Finlay, Ottie and Orla, with my love

Contents

'If fishing interferes with your business, give up your business ... The trout do not rise in Green Wood cemetery.'

Sparse Grey Hackle (Alfred W. Miller),

Fishless Days, Angling Nights (1954)

WITTGENSTEIN'S CORNFLAKES

'I dream every single night that I am fishing.'

Ted Hughes, *Letter* (24 February 1957,
to brother Gerald and wife Joan, in Australia)

Towards the end of his life, the Austrian philosopher Ludwig Wittgenstein subsisted largely on a diet of cornflakes, reasoning that once you have found something which truly agrees with you it makes sense to stick with it.

For the last six decades, fishing has been my cornflakes. But, despite the fact that it has informed and watermarked much of my adult life, oddly enough I have never really thought of it as an obsession.

The origins of fishing for fun – as opposed to food – are inevitably opaque, but it seems to be a practice at least four thousand years old. In seeking out the wellsprings of this idiosyncratic pastime, one must acknowledge that it is a peculiarly individualistic pursuit: *Homo piscatorius* is hard to classify, as everyone fishes in their own different way, and the experience can be as intricate or straightforward as you care to make it. Fishing is such a subjective and protean business that it resists formal definition; once asked by an admirer to explain jazz, Louis Armstrong replied, 'Lady, if ya gotta ask, ya'll never know.'

Even today, when the blandishments of the Great Indoors have never been stronger, angling enjoys a worldwide following and remains a potentially democratic pastime that can be pursued from youth until extreme old age, and it appeals to all types and estates of people who might otherwise appear quite dissimilar. The historical spectrum includes Bing Crosby, Billy Connolly, Robert Redford, Darwin, Eric Clapton, Landseer, Johnny Cash, Oscar Wilde, W. G. Grace and Babe Ruth. Charlie Chaplin was a member of the Catalina Island Tuna Club. George Melly once sold a Magritte to finance his Welsh fishing retreat. After the civil wars, Augustus Caesar took up his rod *animi laxandi causa* (to soothe his spirit) as did Charles II following our own Great Rebellion.

Where fish were concerned, Cleopatra, Nell Gwynne, Coco Chanel and the late Queen Mother were all enthusiastic femmes fatales; in 1922, the record British salmon was landed by Miss Georgina Ballantine (it weighed 64 pounds). One of the first ever best-selling works in English, *A Treatyse of Fysshynge wyth an Angle* (1496), was supposedly written by a mysterious mediaeval prioress, Dame Juliana Berners. (Although the majority of sportfishers are still chaps, hereinafter please let it be understood that terms such as 'angler' and 'fisherman' are intended to embrace all feasible genders.)

There are plenty of other reliable ways of getting hold of fish – poison, spears, explosives, nets, traps, dams, jigs, bow and arrow, snares, trained cormorants and your simple bare hands – but angling properly involves a hook (Old English *angul*, which probably derives from the Sanskrit word for 'bend', *Anka*). The earliest examples of these date back some twenty thousand years, and their invention is arguably as significant to civilisation as the development of the wheel or stirrup. Piscatorial history is a running index of human ingenuity, abounding as it does in abstruse items of tackle, bait and methods, reflecting our devices and desires across a vast geographical range. The artist Turner possessed a rod-cum-umbrella contraption, and once happily spent two days in the rain beside a vicarage pond without a single bite. On Russia's Ponoi river I met a Frenchman – he was sharing his rod with his paramour Elaine, an African American chanteuse, something of a *rara avis* in the Arctic Circle – whose idea of extreme sport was to fish for alligators in the bayous of Louisiana, using as bait a live chicken.

How to catch more fish? Well, wouldn't that be nice to know. This is a devotional rather than an instructional book, so you won't find many practical tips. There is a modern tendency towards specialisation, but I am more of a generalist, intrigued by the soft side of the fishing experience, as opposed to its technical dimension. My personal preference is for fly-casting, where practicable, as I find there is something elegant, harmonious and pleasing about it when properly done (for instance, when I watch others casting) and I have an especial penchant for saltwater 'swoffing' – the Australian term for fly-fishing in the sea. But I will as happily dangle natural baits, sling cutlery, harl, guddle, troll, trawl, trail or leger. While usually I comply with the local rules of engagement, I confess also to a little familiarity with the dark side, and on an army base in Germany was once arrested by the *Fossmeister* for poaching in his stew pond.

Over the decades I suppose I have attained an ordinary level of proficiency in some areas, but there are notably others – bug Latin, engineering-degree knots, taming a centre-pin reel, beach-casting, night fishing for carp – where my skill remains minimal, and, since therefore I cannot truly claim to be an all-rounder, I would follow Wittgenstein's precept, 'Whereof one cannot speak, thereof one must be silent.'

The years have probably taught me to develop a talent for luck, and I like the notion that the more one knows the more one is able to learn. Some anglers are naturally blessed with 'hands' – the equivalent of a gardener's green fingers – but for most of us the ambition must be merely to make as few mistakes as possible, once we have mastered the rudiments of technique (casting is gear change, watercraft resembles road sense). So, although I have clocked up enough experience to qualify as a veteran, perhaps, what I definitely am not is an expert.

Fishing tends to be a great leveller and, if it teaches you anything, maybe it should be humility. The angling world already has its fair share of gurus, swamis, mavens, born-again hunter-gatherers and self-appointed fish whisperers, specialising in oxygen theories, Triple Spey casting methods and recipes for infallible fly patterns – professionals with logos on their hats or whisky on their breath, with all the airs and graces of some love child of Dame Juliana and Papa Hemingway.

Anders Ericsson of Colorado University was the first to suggest that it takes an estimated ten thousand hours of prac-tice to learn how to play a stringed instrument really well, and that's not far off my own experience in respect of a rod strung with fly-line. It may be no coincidence that one of the greatest ever casters, Alexander Grant – 'wizard of the Ness' – designed his Vibration series of greenheart rods along musical principles (he was a consummate fiddle maker, and also cor-responded about mathematics with Einstein). I have certainly put in many hours on becks, brooks, burns, tarns, stanks, lochans, foss pools, chum bars, lagoons, mangrove creeks, tailwaters and permit flats, but I remain an average technician. I catch about as many fish as I deserve, but, partly through good journalistic fortune, I have committed a lot of fishing, and have swum my hooks in almost forty different countries. For those of us who lack unusual talent, there is no substitute for time spent assiduously on the water. Thomas Tod Stoddart, the Victorian author, and sage of Kelso, accounted for 61,573 trout and 928 salmon off the Tweed system alone. He had qualified as a lawyer, but somehow never got round to practising. A former school friend once enquired of him what he was now doing in life: 'Doing? *Doing?*' he thundered. 'Mon, I'm an *angler.*'

You often hear it said – occasionally from those who have enjoyed a blank day – that there is more to fishing than the mere catching of fish (indeed, *piscator non solum piscatur* is the motto of London's august Flyfishers' Club). Some infidels – or 'civilians', as I like to call them – believe there is actually *less* to the pastime than meets the eye.

Angling is a centrifugal activity, spinning off a number of diverse and diverting crafts and skills such as entomology, cooking, taxidermy, photography, meteorology, pisciculture, fly-dressing, and natural history in the broadest sense. Historically, each of these has been represented in a number of related studies, and collectively they make up the etceteras of the sport – a kind of cabinet of curiosities. Many of the mainstream authors I most admire for their panache – Rabelais, Burton, Swift, Sterne and, in the vast *bibliotheca piscatoria* itself, Walton and his disciples – are digressive by nature, and I have approached this subject in similar fashion. Some anglers are quite happy to concentrate merely on the practical aspects of the subject, of course – but whether you stick to the missionary position, or riffle through the *Kama Sutra*, it's still an act of love.

I am also convinced some subtle enchantment is involved in the appeal fishing makes to all the senses. I have a tin ear for music (even for the 'Trout' Quintet), but I relish the varied soundtrack of the waterside – the sudden crash of a sea trout in the gloaming 'when the green has gone out of the grass', gulls screaming over an offshore baitball, the corncrake rasp of your venerable reel as a deep springer turns off his lie and feels the steel. I am transported by sea music, ocean whisper, the busy clack of palm fronds in a rising breeze. Then there is the velutinous feel of a new flor-grade cork handle, the first twitch on the line when touch legering, the pleasing grasp of a salmon's tail wrist in your hand.

Visually, the fireside angler reruns with contentment the spectacle of summer grilse glancing across the tail of a run, a mangrove flat at dawn pulsing with tarpon shiver, the reverse blizzard of hatching mayfly. One savours the sight of that belly-button rise beneath the far willows, which momentarily becomes the *omphalos* of your immediate universe. Before releasing him, you admire the sculpted titanium head of a bonefish, his overall radiance, his neatly underslung mouth like a Morningside auntie cooling her cock-a-leekie. Elsewhere, there are evocative olfactory notes of watermint, turfsmoke, wild garlic, the lavender scent of sparling, the sharp whiff of bankside whisky from your slip cup, the lanolin snuff of damp tweeds. As another jazzman said, 'Brother, if that don't light up your life, you ain't got no buttons!'

In the realm of taste, however, it seems our quarry has evolved a hugely refined superiority: some fish can detect chemicals (such as seal oil) in dilutions so minute as to be the equivalent of a single drop of vermouth in 500,000 barrels of gin. Jeez, now that's what I call a Martini *con cojones*.

One reason it is hard to work out the algebra of such a complex passion is that fishing – the epitome of wishful thinking – seems to offer paradoxical sensations. How can it be simultaneously sedative and tonic, concentrated and relaxed, footling and sublime, exquisite yet melancholy? Additionally, there is the amalgam of finesse and ferocity, a volatile sequence of voluptuous serenity followed by batshit frenzy and alarm. While it can be a speculative occupation (in the same sense that panning for gold might be), it doesn't seem much like the contemplative man's recreation when your boat is in the middle of a magimix school of bluefish, or a late-evening rise on a hill loch when all the midges in Caledonia decide to embrace you.

On occasions, there is a bittersweet aspect to it, and angling may be laminated by despair as well as delight. But fundamentally it concerns the pursuit of happiness; some critics claim this is an emotion that 'writes white' (as opposed to colourful and promotable misery), while I prefer to think of that Venetian sundial inscribed, *'Horas non numero nisi serenas'* (I don't record the hours unless they are cloudless). Ever since my first, close encounters with fish, the resultant pleasure has become dissolved into my entire life, commingled like the milky furls of an antiseptic poured into warm, childhood bathwater. This, therefore, is on the whole a chronicle of contentment, and I am unashamedly a celebrant of its diversity.

Setting aside the multitude of books that have been published specifically about fishing, it has attracted an unusually large number of literary writers – Shakespeare, Ben Jonson, Andrew Marvell, Jonathan Swift, Turgenev, Tolstoy, Kipling, Chekhov, J. M. Barrie, Yeats, Robert Lowell, Ted Hughes, Tom Stoppard. This may be because in some lights it can become a mythopoeic activity, suggestive of other aspects of life – in its strata of challenges and tribulations, its perpetual illusion of treasure trove shaded by the spectre of disappointment – and something capable of transforming the mundane into the fabulous. I certainly sense there may be an affinity between the precise casting of a line and the exact crafting of a line of words – and if you don't believe me, you can away and boil your head for soup.

Ernest Hemingway, who was introduced to big game fishing by the novelist John Dos Passos (he used to mix their onboard cocktails in a zinc pail), grew so impatient playing Gulf Stream sharks that he occasionally dispatched them with a pistol, once splintering his own knee. Arthur Ransome – who liked to describe his politics as 'Fishing', though he was married to Trotsky's secretary – claimed the creative act of which he was most proud was devising the Blue Vulturine Elver fly (there is a nice portrait in the Garrick Club bar of him tying one at his desk). In Oxford, Mississippi, I was invited to fish the migratory white perch run using one of William Faulkner's trusty rods.

It is an activity much concerned with excitement, too. Orwell (another angler) has his protagonist George Bowling, the insurance clerk in *Coming Up for Air* (1939), exclaim, 'The thought of fishing sent me wild with excitement,' which pretty much sums up what is known as 'fin fever'. This is a sensation that effervesces throughout your bloodstream, crackles the synapses, palsies the limbs, renders the mouth dry and inclines your heart to lodge against its ribcage. It is caused by the glorious and visible prospect of an approaching fish, especially if you are armed and ready. It is one of the key reasons aficionados describe angling as addictive – this, and the subsequent moment of the take, comprise the fix. (There *is* more to fishing than just catching fish, just as there is more to climbing than the mere bagging of a peak, but it remains the chief purpose of the enterprise.)

Ignoring shark schlock clichés for the time being, fin fever is the quintessence of promise, of imminence. A classic instance is the beguiling, leisurely 'head and tail' rise of a fresh salmon porpoising with that slight curling motion in the head of a run (one of the more gorgeous sights in a fly-fisher's world). A shoal of bonefish scissoring the morning flats like satin, or the dorsal of a great, solitary carp beyond the reeds will make most hands shake in similar fashion. One of the more spectacular moments I recall was in October 2006 when we were experimentally trolling with fly rods off an inflatable dinghy alongside an uninhabited atoll in the southern Seychelles and a large blue marlin – all lit up with neon stigmata, his quivering sail erect – surfaced maybe 40 feet behind the boat, lunged, and was hooked. He vaulted several times, as if he would tear a hole in the sky – but I was more concerned that his bill would tear a hole in our flimsy craft (there were several hefty tiger sharks also cruising the drop-off). When you are squatting on an air-filled rubber ledge that low to the water, a leaping fish of several hundred pounds is actually performing its aerial lambada above you. Fortunately, after twenty minutes, our Man in the Blue Suit threw the hook.

There is something iconic and universally recognisable about a fin breaking the surface, and it has been used as an image for the *moment critique* of literary inspiration by both Marcel Proust and Virginia Woolf. The former likened a rising fish to the instant wherein you discover the precise configuring of a phrase – a welcome discovery arriving from the subconscious depths: fittingly for the author of such a mighty *roman-fleuve*, he harboured fond memories of fishing as a boy, and even compared the layers of memory to a childhood creel of still-living fish kept fresh under a covering of cool grass. Woolf – who was alive to the mystique and resonance of angling and, rather surprisingly, wrote an essay on the subject ('Fishing', published posthumously) – imbued much of her work with aquatic symbolism, perhaps aptly for a writer who was exploring so-called 'stream of consciousness'. In her *Diary* (30 September 1926) she describes a fruitful scanning of the imagination as 'one sees a fin passing far out', and then, triumphantly on 7 February 1931, having just completed her novel *The Waves*, 'I have netted that fin in the waste of water.' Ten years later, with her overcoat pockets full of stones, she went to her watery death in the River Ouse.

*

I'm not sure Wittgenstein was ever interested in fishing, though he was quirky and colourful enough to have been an angler (civilians might claim he was altogether too sensible to be taken in by anything so fatuous). Briefly at school with Hitler, he inherited a steel fortune, designed propellers, pioneered logical atomism, refused to wear neckties, once threatened Karl Popper with a red-hot poker, and eventually resigned his Cambridge post before retreating to an Irish fishing village, where he befriended a seagull. Nonetheless, one of his strictures about religious belief strikes me as especially relevant to the angler's enquiring mind: 'to know that the facts of the world are not the end of the matter'.

To the uninitiated, ours does not appear a strictly rational pastime (the French have a phrase for eccentrics as being *capable de pêcher à la ligne*), and there have been sufficient examples of zealotry that teetered over into the type of anglimania beloved of cartoonists – for instance, the classic *Punch* lampoon of a Dottyville inmate beckoning to an angler over the asylum wall, 'Come inside.' An eighteenth-century librarian, the aptly named Thomas Birch, used to disguise himself as a fully branched tree when setting off for the water, and was eventually taken into care. The deranged Philip V of Spain was confined to quarters and for years fished for a single carp freshly stocked each day for him in a bucket (similar to the gouty angler and his washtub depicted in Theodore Lane's celebrated picture *The Enthusiast*, now in the Tate). Sir Humphry Davy, who 'abominated gravy' and discovered sodium, was a frequent guest of Walter Scott and would turn up dressed like a Dutch pirate, with porpoise hide hip waders spattered in salmon blood.

In 1995, Peter Eastman Jr, a Californian teenager wanting a fancy moniker for his high school graduation certificate, paid $192 to change his name by the equivalent of deed poll to 'Trout Fishing in America' in homage to Richard Brautigan's 1967 counter-cultural novel – the only work of fiction I know that ends with the (deliberately misspelled) word 'mayonaise'. His father expressed regret that their family name was being lost to a fish. Extreme devotion to trout may even fatally have affected the course of history. As James Owen suggests (*Trout*, 2012), the tempting abundance of fish in Wyoming's Tongue River so preoccupied the troops of General Crook's army that it delayed their march to relieve Custer and the 7th Cavalry at Little Bighorn. According to Captain J. G. Bourke's meticulous account (*On the Border with Crook*, 1891) they consumed some fifteen thousand trout in three weeks. (Bourke was also the author of a very different survey, *Scatalogic Rites of All Nations*, that concludes with instructions for the use of bladders in making excrement sausages. But, enough of madness.)

By contrast, there is a long history of detractors depicting the pastime as stultifying and absurd. Plutarch accounted it 'a filthy, base, illiberal employment, having neither wit nor perspicacity in it'. Christ on a stick! The apocryphal aphorism from Dr Johnson about a stick with a worm at one end and a fool at the other properly belongs to the poet Martial Guyet from the seventeenth century; in fact, the good Doctor approved of fishing, and could quote passages verbatim from

The Compleat Angler, which he referred to as 'a mighty pretty book'. But it is the charge of cruelty, rather than folly, that has been most pronounced.

In a magnificent display of hypocrisy – given that, in an 1814 letter, he had recorded, 'I have caught a great many perch, and some carp' – the poet Byron noted in *Don Juan* that it was 'the cruelest, the coldest, and the stupidest of pretended sports', concluding, 'No angler can be a good man.' There are some who would agree with him, but although I must concede that on occasions the fish may be discomfited by our interference with their apparently innocent lives, the angler takes his pleasure and finds fulfilment not because of this, but despite it; and I believe it to be true that you cannot be an authentic hunter (fishing being, in Robert Burton's phrase, a 'hunt by water') without feeling a certain love for your quarry – which is by no means the same thing as bloodlust. Although it is considered barbaric by objectors, the fact is that the hunter may lay siege to an individual creature but he tends to be a friend to the species as a whole.

I would also suggest that angling is more to do with the life, as opposed to the execution, of fish. It is unique among field sports in that there is usually the option of release – hookset and trigger squeeze may both constitute assaults on unsuspecting creatures, but the latter, when successful, is irreversible. There is increasingly a belief that you should let your fish go, because it deserves it (sometimes called 'kiss and release'), but not all subscribe to this view. Hemingway was famously keen on wildlife slaughter, and Brigid Brophy, the spirited novelist who also wrote widely about animal rights, had an especial detestation for what she saw as that Nobel laureate's posturing machismo and atavism, dubbing him H. E. Mingway, and crisply remarking, 'His chum Gertrude Stein was three times the writer, as well as twice the man.'

Byron also onanistically alluded to angling as 'that solitary vice', and this phallic innuendo was taken up by another opponent, Maureen Duffy, who asserted that all forms of hunting were effectively masturbation fantasies, and holding up trophies was part of 'the whole penis folklore'. I have even heard it said by certain female anglers that playing a fish on a rod is the closest thing they can experience to an erection. Is it conceivable there is indeed something priapic about all of this – an element of lust underlying that urgent desire to achieve gratification? There is undeniably a fishy dimension to sex, and possibly vice versa (in Pompeii, a place with many love shacks, Cupid and Venus were depicted as fishing together). Even a cursory trawl through a lexicon of historic slang will show the figurative association of fish with a carnal and lascivious range, including sodomy, underage intercourse and the pudenda. The Micronesian language Yapese is not alone in having a word for clitoris that means 'small fish' – *qathean*, for any of you college professors out there.

Shakespeare (who was arguably no more bawdy than any other Renaissance playwright) has Pompey Bum, Mistress Overdone's barman-cum-pimp, refer to adultery as, 'groping for trouts, in a peculiar river' (*Measure for Measure* I. ii. 83), and Little Tom Tittlemouse of nursery-rhyme fame was equally out to cuckold when 'he caught fishes in other men's ditches'. The identification of fish with venery actually derives from the cult of Venus/Aphrodite, coupled with the worship of that most fecundative of fish, the carp (named cyprinids, from Cyprus, the mythical birthplace of the goddess), which gives us eventually *dies Veneris* – thus, fish on Friday. There is also an elaborate global history of fish-related aphrodisia, and Catullus informs us that the punishment for adultery in ancient Rome was to have a mullet thrust up your fundament (he does not specify whether red or grey; but, either way, being a purveyor of Mugiliformes must have offered fairly steady employment).

In the 1990s, a survey of North American anglers (in this case almost exclusively male, I imagine) recorded that 40 per cent of respondents definitely preferred fishing to sex. And there was me, casting away for several decades without ever realising there was even that choice.

It used to be argued that fishing was an honest diversion associated with thrifty monks and clerics, an occupation sanctified by scripture: 'the sport and Recreation of God's Saints' enthused Gervase Markham back in 1614, though the Zebedee family and other disciples were in fact professional netsmen. Izaak Walton identified angling with the dispossessed Anglican community during the rule of the Cromwellian saints, and it would be hard to disagree with the rector in George Eliot's *Middlemarch* (1871) who observed of the learned Casaubon, 'It is a very good quality in a man to have a trout stream.' But, might there actually be qualities that one could call Angling Virtues?

The proverbial 'patience' – though habitually assumed – would certainly not be one of them. It is a profound misconception that anglers are gently submissive folk who pass most of their time dreamily static at the water's edge, blinking occasionally like livestock at a ford. I'm sure I'm not alone in being volcanically *impatient* (fin fever will do that to you), and many sportsmen of my acquaintance are quietly subversive by temperament, constantly on the lookout for ways of turning things to their advantage. But if I were to list the angler's desiderata I would hymn the virtues of 'persistence' first and foremost. W. C. Fields's philosophy, 'If at first you don't succeed, try, try again. Then quit. There's no point in being a damn fool about it', amusing though it is, will never help you to Catch More Fish.

Piscatorial history furnishes numerous exemplars of this dogged determination. Despite losing his right arm to grapeshot at Tenerife, Nelson continued his fishing life single-handed (on the Wandle, among other streams); one can only admire his remark to a similarly wounded sailor, looking down at his own pinned and empty sleeve, 'Well, my poor Jack, then we are both spoiled for fishermen.' His indefatigable contemporary Colonel Hawker of Longparish, whose hip was shattered by musketry at Talavera, took to fly-casting from horseback, and accounted for twelve thousand fish from the Test alone (he was also a prodigious shot, once pulling off a right-and-left at a bat and a stag beetle;

which makes him a good marksman but not, I concede, necessarily a good man). And I have long had a soft spot for Tom Stoddart's stalwart friend 'Timbertoes', who had three wooden legs – when wading in Tweed in cold conditions, he used the third as a tripod, 'and sat doon as cozy as if he were in the chimney neuk'.

Along with an unwavering stoicism, the angler needs passion, optimism, dexterity, honesty, generosity, a capacity for lateral thinking, humility, curiosity, good manners, the ability to improvise, tidiness, compassion and good grace (I could try tacking on sobriety, but my heart isn't in it). Any apologia for the sport, however, has also got to tackle the aura of mendacity that has surrounded it for centuries.

I don't believe anglers are inclined to fib or exaggerate more than anyone else, though admittedly the literature abounds in tall tales to a notable degree. Is it any coincidence, I wonder, that it has been the *passe-temps* of choice for so many statesmen and politicians? Lenin was an enthusiastic ice-fisher, and Vladimir Putin makes a considerable show of his stage-managed prowess with rod and line. General Franco had a soft spot for salmon angling. Nero cheated by using nets of gold thread, though Lear's Fool informs us he was posthumously 'an angler in the lake of darkness'. For presidents of the USA, it has been almost a mandatory hobby: Hoover published a book on the subject, FDR overcame physical disability to pursue billfish (and was also fond of harpooning manta rays), Carter kept a fly-dressing desk in the White House, and I once spotted George Bush *père* in the Florida Keys, his strike rate not being materially assisted by the skiffload of security men spooking the flats.

It was George Washington, however, who initiated this tradition. As a kid, he broke his papa's precious fishing pole – and when questioned about it later, in true angler's fashion he replied: 'Father, I cannot tell a lie. It was Jerry Perkins from next door that did it.'

The trait of paramount importance, however – and this becomes particularly germane when choosing your fishing buddy – is a sense of gallows humour, and the cultivation of a steely strain of irony. This is not the same thing as finding the pastime intrinsically comical, but it can be vital to recognise those inevitable moments of absurdity when the spectacular failure of your best and most earnest plans only serves to highlight the helplessness of our human condition under the eye of pitiless deities. You will particularly require the balm afforded by a fatalistic acceptance of cosmic injustice when you suffer a blank day, as then the greatest recordable distance in the universe appears to be between No Fish and One Fish – the difference between one and two being infinitesimal by comparison.

Imagine a situation in which misfortune to you is married with success to others (for most of us, this does not require a great feat of mental fabrication). Your net is dry, you have just lost the only fish of the day, and naturally it was a nice one. The cause was your own negligence – a hastily finished knot that slipped – and you are just inspecting furiously that hateful little piggy-tail of unravelled nylon, contemplating the foolish loss of your heart's desire, and considering an early withdrawal to the bankside bothy where you will work on a bottle of pre-luncheon gin, when over the bridge jauntily saunters your former mate, wader fronts glistening with slime, wearing a grin like the second sunrise over Eden, and lugging a fish-bass that bulges obscenely like some Jacobean codpiece; and, as the perfidious gods of the Angle snigger amid the cow parsley, he hails you with a merry catcall, commiserates on your loss (all of which he witnessed in humiliating detail) and with histrionic effort hoists aloft his sack of several trophies. Now, I say, you must ignore that acidic seep of envy familiar to us all, and resist the accompanying impulse to seize his wretched Loomis rod and use it upon him as a proctologist's speculum, for your dignity depends on equanimity in the face of the gross provocations of the cosmos, and you must not give in to Rod Rage (which requires Angler Management), remembering instead the response taught us by St Polycarp, patron saint of the jinxed, the skunked and the benighted. The words you are looking for, I believe, are, 'Ah, well done you, matey.'

Angling is predicated on uncertainty. We are forever chasing shadows, and much of the time you cannot be sure there's even a fish out there. It's a little like a seance – 'Is there anybody there?' You must have the courage of your convictions, and be a firm believer. I think of the process as an act of faith launching a confidence trick – your fooling of the fish, your broadcasting of fake news (take, and eat). And although offstream I am a pessimist of some conviction, when I go fishing I have to believe anything is possible, including the semi-miraculous, and conceivably even an act of God.

The great Norman Maclean – whose 1976 novella *A River Runs Through It* was a landmark in piscatory fiction (despite being turned down by one publisher because 'These stories have trees in them') – was also a distinguished English professor at the University of Chicago, whose students included Saul Bellow and Philip Roth. Each year, he devoted the whole of his introductory class to discussing the first line of the world's most famous play, 'Who's there?' (*Hamlet*), on the basis that the entire drama is an answer to that opening question (the ghost is about to put in another appearance). It could serve as a motto for the sport of fishing, which is essentially an interrogative process – a quest that sometimes seems to be asking for the impossible. Even the hook itself resembles an inverted question mark, like the Spanish one.

Because, despite the millions of hours invested in schemes for the downfall of fish, there is still so much we simply don't know: how do swordfish navigate; why *do* salmon occasionally intercept a bait? These beautiful mysteries are surely part of the allure, haunting the margins of the pastime like will-o'-the-wisps over marshland. I relish that my sport remains a potent cocktail of voodoo with just a jigger of science admixed, and for every practical consideration about relative air/water temperatures, or the exact sinking rates of a Skagit tip, I favour a nice splosh of wizardry and superstition. It remains beguiling precisely because we cannot truly parse it.

If there is something magical about angling it may be because at its heart lies a basic conjuring trick – producing a fish out of water (with wing-shooting, for instance, there is an act of reverse prestidigitation, whereby the life of a bird is plucked from the air and disappears into the hand). The prestige accruing to the practitioner depends on this sleight of hand, but also upon a degree of luck – something we try to maximise in various ways, developing a positive mindset, believing in our powers, and conceivably even entering a state of grace. Of course, fishing being a broad church, you could just clutch your roach pole and hope for the best. You are going to be spellbound, either way.

Angling is also distinguished from other hunter-gatherer activities by the fact that there is a temporary collusion between you and the quarry, and your life is connected directly, down the line, with the life of a fish – and, by extension, with the whole of the natural world. It becomes your hotline to the universe, a tenuous copula that establishes a circuit of power, the source of vigour and refreshment and energy, like some lightning conductor (I call this the 'lightning thread', a phrase borrowed from Dylan Thomas's apocalyptic saga 'Ballad of the Long-Legged Bait', 1941). And I'm convinced it is no coincidence that, at this moment of connection, the battle curve of your rod, the straight length of line, and the corresponding curve of the hook describe, in their serpentine configuration, Hogarth's 'Line of Beauty'.

Thomas Hardy wrote an eerie poem, 'The Convergence of the Twain', about the inexorable coming together of the *Titanic* with its iceberg, and I feel something similar (especially with migratory species) when my life and the life of a fish converge and intersect – one should not entirely rule out the possibility that the fish is awaiting you, too.

Sometimes, all things conspire, and the multiverse seems to be in alignment – there's a sudden streak of light under the water, and there comes what bullfighters call a 'moment of truth' (*momento de la verdad*). Precisely what truth is revealed will depend on you – it could be a Joycean epiphany, or just the realisation that changing down a fly size suits a falling water. The coming together that is a 'take' produces that little starburst of delight, the biochemical click, which is an integral part of the angling fix. Norman Maclean invoked Wordsworth's 'spots of time' – 'eternity compressed into a moment' – but I know of no better description for this sublime experience of harmonious convergence than a phrase coined by an Irish gillie and recorded by A. A. Luce, an eminent Irish professor of metaphysics, in his remarkable book *Fishing and Thinking* (1959): 'the fishological moment'.

Now well past sixty, and having felt the breath of the Pale Rider upon my nape, I like to think I have taken to heart the advice of that remarkable American author 'Sparse Grey Hackle', which forms the epigraph to this book. I make the most of every opportunity to go fishing, and I have even visited a Greenwood cemetery (in Mississippi) to confirm there were no fish rising there. A Saxon bard once compared our fleeting lifespan with a swallow swooping briefly through a

23

brightly lit mead hall, then rapidly disappearing back into the night. Until I go to my long home, I daily pray to the Lord God (Wildlife Department) that I may continue to search – rod in hand – for Paradise on earth. It is a pleasurable quest: a man might hope to discover something that was not without meaning, if never exactly a Golden Fleece, or the Lost Chord. And perhaps the facts of the world are not the end of the matter, it's hard to tell. Is there some truth in the assertion, attributed to H. D. Thoreau, that 'many men go fishing all their lives without knowing it is not fish they are after'? I couldn't really say; I have only been at it for about sixty years, so am still trying to get the hang of it.

2

THE HOUSE OF TONGUE

'Early on, I decided that fishing would be my way of
looking at the world.'

Thomas McGuane, *The Longest Silence* (1999)

In the beginning was the worm.

We have never before been fishing together, and my father
knows virtually nothing about the pastime when my yellow-
handled Japanese bait rod is fixed between the spongy clips
on the gutters of the field grey Land Rover's roof and we
set forth for the nameless burn. It is August 1963, and I am
seven – a pale, quiet, somewhat daydreamy boy, eager to
please, perhaps a little solitary, an only child, bookish, urban
and pampered, wearing an oversized green wind-breaker and
a borrowed tweed cap.

It had not been the happiest of years for our family. In June, my father had resigned from Harold Macmillan's government during a very public furore, and since then my parents had been virtually in hiding – shunned by many of their former friends, pariahs of their social and political circles, generally considered to be in disgrace. As a way of avoiding the lime-light, they had rented for several months a remote lodge on the Sutherland coast. Overlooking a moody kyle (or fjord), its gardens surrounded by a high wall that kept out the pack of journalists, the House of Tongue was our temporary refuge from the outside world. This was the year when, according to Philip Larkin's poem, sexual intercourse had been invented; but, in our Profumo family enclave, with its comforting peat fires and formal mealtimes, one might have been forgiven for thinking the Sixties had never begun to swing.

Its stubby wipers make a reassuring thud as the car slurs through puddles on the moorland road. We are not being fol-lowed: a diversionary tactic has drawn away the attention of the photographers loitering by the locked drive gates. Neither of us has much idea what lies ahead, but I'm pleased it is just the two of us – a rare enough occurrence.

'Do you think we'll catch any?'

He waggles his forefinger diplomatically. 'You can never tell, General' (his nickname for me when he was Secretary of State for War). I imagine he has other things on his mind.

My father is forty-eight, and quite good-looking in a dark, Italianate way (some of his Westminster detractors dubbed him 'The Head Waiter'). He is wearing thick beige gardening cords held up with felt braces, and his favourite tweed trilby with a jay's feather in the band. I doubt he'd ever shot a jay, so the hat must have come from Lock's complete with this embellishment.

Much of the previous night the rain had been rapping its knuckles on the corrugated roof of the outbuildings, chirruping along the rones – but this afternoon it is dwindling, and the air is close, though there are still caviare skies ahead. On my lap is a St Bruno tobacco tin seething with worms.

The House of Tongue was once home to the Mackay chieftains – their motto *Manu Forti* (with a strong hand) was inscribed on a pediment in the dining room, now darkly carpeted with the tartan of the Sutherland clan, whose estates in their heyday had stretched from one coast of Scotland to another. In his striped trousers and new Chelsea boots, Mr White (my uncle's flamboyant butler) attended to meals for the grown-ups, and the kitchen was run by Mrs Bivar, a lizard-skinned Anglo-Indian widow with a sweet, deep burr of a voice. She was spectacularly bent from childhood polio, and her twin signature dishes were mulligatawny soup and crème brûlée, both of which were served up to me and Nanny Burch in the nursery. Small wonder my elder half-brother Mark liked to call me 'Little Lord Fauntleroy'. I was blissfully insulated from the iniquities of the wider world.

Naturally, there was a ghost – a smiling, grey lady, only the upper half of whom was ever visible as she crossed the attic rooms where, in Victorian times, the floor had existed at a lower level. I never saw her, but I was properly spooked by the green shore crab Miss Burch brought back from her walk one morning and carelessly let loose into my room, where it scuttled beneath the bed. I have ever since dreaded these creatures, with their bubbly mouthparts and geriatric frowns, that sideways slink, and the way they gesticulate like stock-jobbers. I concede there are worse childhood traumas, but it has made difficult a lifetime of summers spent dealing with lobster pots.

The idea of renting the house had come from my venerable uncle (Harold Balfour), who assumed control of proceedings. Despite his benevolence to me, I went in awe of him from the start. An Edwardian sporting gentleman of formidable aspect, he was profoundly deaf from having spent years flying open-cockpit aircraft during the Great War, during which he had been awarded the MC and bar. Aged just twenty, and still technically a minor, he had been promoted to the rank of major in the Royal Flying Corps, and survived a dogfight with the Red Baron. He used to make me little gliders from folded card, their noses balanced with a paper clip, and treated me like a grandson. I didn't then know how loyally he had defended my parents in public, when precious few were inclined to be associated with them, but he had become my hero anyway.

As was then the British way, the family pretended with imperial sangfroid that nothing untoward was happening in the world outside, despite the clamour and flashbulbs that accompanied each expedition. A man who had been shot down as a teenager over Vimy Ridge was not going to let a mob of reporters interrupt his plans, so Uncle Harold studiedly ignored them as he led the morning motorcade up to the moors where, properly attired in tweeds and ties, whatever the temperature, the men of Tongue would pursue grouse over insubordinate setters. Donny Murray was the keeper – not one of those taciturn retainers kippered by years of Glenfiddich and disdain, but a kindly, reserved man in waistcoat and watch chain, with a clipped moustache and thick, horn-rimmed spectacles. I enjoyed watching the adults in their rituals before a shoot, but I was never invited out with them.

I was never allowed into the gun room unaccompanied, either, but was dreamily drawn to its heady aromas of lubricating oil, Havana smoke and midge repellent. There were odd souvenirs to be filched, also: a stray wooden rod-stopper, crumbly squares of brown Kendal mint cake, an empty Mucilin floatant bottle with its pleasing cork discus. I collected spent paper cartridges, with their spicy reek, and the clear paper envelopes that had contained knotted casts. When the massive wooden salmon rods were loaded onto the cars, I emulated these mighty preparations by removing a couple of plastic jousting lances from my Swoppet model knights and poking them out of the back window of a Dinky truck. I suppose I was already quite a peculiar child.

The summer we almost burned a castle down was the only previous time I had held a proper rod in my hand. An irregular, pocket-sized Jacobean structure on the banks of the Aberdeenshire Dee, Braemar Castle had once been tenanted by Robert Louis Stevenson, that most accomplished of swashbuckling authors, who also invented the sleeping bag, and wrote much of *Treasure Island* at Braemar during the summer of 1881 to amuse his stepson. One of their neighbours was a certain John Silver.

Aged five, I was under the care of a dyspeptic Belgian governess named Vévé – glaucous eyed, lightly moustached, with a penchant for small funereal hats and a rough way of wiping down the corners of my mouth with her spit-sodden hankie. I was dispatched to the nearby Sluggan Burn one July morning, where somehow, using basic worm tackle, I derricked from the dark water a small bootlace eel, that was eventually wrapped, alive, in the hood of my discarded waterproof before being decanted into the basin of the downstairs cloakroom, where I imagined it might be admired by all, though no adulatory mention of it was made, and next morning it had disappeared – flushed away, perhaps, by the revolted daily lady (migratory eels are historically reviled in the Highlands). Whether somehow provoked by having had to deal with my first ever catch, or through a long-accumulated professional dislike of young children, Vévé decided at breakfast to thrust my Fauntleroy face down into a bowl of piping-hot porridge, for which act of pastoral attention she was dismissed, though I did develop a taste for Scotland, nonetheless.

Before he went up to bed that evening, Michael, the student helping out in the kitchen, raked some hot coals into a pail which, being an antique one made of leather, turned to flame during the night, and the castle caught fire. Several brigades were summoned, but the water pressure was so weak that the blaze had to be treated with (metal) buckets being passed by hand up the circular stone staircase. I was carried out in a blanket, and the only fatality was my aunt's asthmatic Pekinese, who refused to be rescued from beneath a four-poster. My early visits to the Highlands were not entirely serene.

Children don't usually appreciate scenery, but as we rattle along the single-track hill road my father indicates the majesty of the landscape, the vista of shoreline and skerry, the bulging bosses of boulder and crag. It seems quite empty to me. I have been enthralled by *The Chronicles of Narnia*, and am disappointed there are scarcely any animals to be seen. 'Rather wonderful, though,' he suggests distractedly.

We pull into a passing place and peek over the little road bridge at the spating burn that is swishing beneath it, turbid and peat stained, its reedy margins dotted with twirling, caramel spume. This appears to be the place. I am dismally armed for the fray. The plastic rod has been rigged by Mr Murray with a terminal tackle comprising a small salmon iron, a twist of lead wire, an orange-painted cork bung that would have buoyed up a coal scuttle, and a tinny reel packed with heavy brown nylon line. If anything should break or require replacing, we will be in trouble because neither of us knows the requisite knots. We march 20 yards straight up the heathery right bank, not even standing back from the water, and then take up positions opposite a minor bay where the tawny water moils like mulligatawny.

Received wisdom has caused us to scour our worms in sphagnum moss, and already part of the fun has been to fork them out from a mulch heap. From the glutinous container two gilt tails are wrestled messily onto the hook, which has nylon bristles lashed to the shank to secure the bait. My father inspects his handiwork doubtfully, stands behind me, pulls a little line off against the ratchet, and, his arms around me, together we ineptly lob out our offering to the unfamiliar stream.

Several times the bobber dawdles down the viscous current, artlessly dragging its load; then it begins to stammer and twindle. We do not read the telltale signs at first, but when the float jabs sideways, there is that sensation you get when the bath plug you are dangling on its chain is sucked down into the vortex of the draining water. I execute an extravagant strike and I feel the pulsing resistance of my first hooked trout. After a few figure-of-eight wriggles it is hauled, splattering, across the surface and swung way up onto the mossy bank, where my father deals it a couple of rather gingerly blows to the head from the hefty stag's horn priest with which he has been issued. The brownie flickers its pectorals awhile, like the eyelids of some dreaming dog, and then it is over, and it has all begun.

Some memories are blurred askew by the passage of time, and the fancy of one's imagination, but this one remains as clear as spun sugar. We both peer almost incredulously at these 6 ounces of moorland fish, its back the umber hue of ling heather stalks, its flanks gleaming like treacle over a spoon. There are splotches along the lateral line as if raindrops had hit the khaki of summer's earlier dust, and, surviving even the gunmetal sheen that creeps over it in death, the body is prinked with points of rose madder. It was a lean and hungry specimen, but to me entirely beautiful. Gently, I poke the puckering film of its eye.

Surely relieved by this sudden success, my father hugs me to his corduroy lap. 'Well done you, General,' he laughs, and shares his wrapped tablet of dark hunting chocolate, which I dutifully crunch despite the ghastly bitter taste. He takes a good pull from the miniature Bell's bottle in his pocket, making exaggerated gurgling noises like some stage drunk. The rain has passed, and midges are thronging our heads with their blood song; we varnish our faces with citronella oil from the pharmacist's vials, but nothing can impair a scene so splashed and lit with pleasure.

'Let's see about getting another.' He re-baits, and before long the process is repeated – except it will never quite be the same again, not once that experience has bedded in and settled. Parched – fevered – by all the excitement, before we leave I squat by that generous waterway and take a draft of pure hill water. It feels cold enough to hurt like a knife; but something glorious has also entered my heart.

What is it about fishing and childhood? Anglers tend to get a little breathless recalling their early experiences, and talk of the passing of innocence, and a need to recapture that sense of wonder. Nostalgia is the ground elder of memoir, but time and again in the literature one comes across this gentle, insistent longing for simpler, clearer days (let's leave the proverbial bent pin out of it) before adult preoccupations sullied our vision of the world, and dulled everything with useless white noise. Memory Lane can prove to be a tedious little cul-de-sac, but I think that for many of us there is a profound association of fishing with happiness, and that it is one of those rare activities one can enjoy at effectively any age, and that this continuity can prove powerfully reassuring. It's like eating the comfort food that your mama used to cook – at least, so I fondly imagine: mine could scarcely make toast.

One of my favourite passages that evokes this sharp sensation of retrospect comes from Lord Grey of Fallodon's classic *Fly Fishing* (1930 edition). The former foreign secretary, having declared 'the lamps are going out all over Europe' as the nations sloped towards the war, had lost two wives and two brothers, both his homes had burned down, and the level of his eyesight was sinking fast, yet there is nothing maudlin as he delights in the boyhood sessions he spent on Scottish burns with an elderly keeper who liked to take it easy with his pipe: 'When I now recall that distant land, I see always somewhere among the whin bushes a little curl of thin smoke.'

Golden Ages have a habit of receding. It's not so much that one wants to brood over the triumphs of far-off days (the remembrance of fish past), though that does have its attractions when you have your feet up on a warm fender in February, perhaps. Rather, there is contentment merely in knowing that a relatively smooth connection exists with a discrete set of personal episodes, despite the various snags, logjams and other impediments that the rest of your life may have had to bounce and career over, causing the usual but unpredictable turbulence. Whatever the vicissitudes of your non-angling life, here you always feel close to the regular heartbeat of something unassailable.

Indeed, there can be a pronounced other-worldliness about the dimension you enter while fishing, and I think it has a fairy-tale quality that again resonates with the imaginative shapes of childhood – the mounting of an expedition, the central act of deception, the capture of an animal, that general penumbra of magic, the rituals and convergence, and rough justice. These are motifs of considerable potency, and for the fisherman I believe this is focused by our enduring fascination with water, that mysterious, reflective element that both separates us from, and offers a peephole into, some parallel world (not that one would ever have time to give such matters a second thought when actually out enjoying oneself at the water's edge).

Setting aside such cosmic speculation, you should never forget that early experience of fishing as fun. 'Fishing simply sent me out of my mind,' wrote the Russian author Sergei Aksakov. 'I could neither think nor talk of anything else.' Though not with quite that degree of delirium, my own boyhood was similarly transformed, and what began as a pleasant background hum in due course became the soundtrack to my entire adult life.

Perhaps I should never have taken that first drink.

In the entrance lobby to the hotel at Tongue, where we sometimes went for high tea, there was an immense cased salmon – a crinkly-skin mount with gold lettering on its bow-fronted glass that recorded the weight as 32 pounds. It had been taken one spring during the Great War from a bouldery local river where my uncle had arranged access to a beat every week.

The riparian owner was a tall, watery-eyed major (there was a plethora of them in the early 1960s) who had retired to live quietly on its banks, which he patrolled with his thumb-stick, dressed in regimental tie and an alarmingly stained brown trilby. He was known to be thirsty, but I never saw him drinking. Though he doted on his little river, he had given up fishing it himself – 'Doesn't give me a cockstand any more,' was his explanation one afternoon to my astonished aunt, a sharp-faced lady with a habitual twin-set and packet of Du Maurier cigarettes. Whenever I joined the others for a bothy picnic, he would ignore me, as befitted my lowly rank, though to adults he was both charming and shy. When the water was low and unproductive, he would commiserate: 'It's a fair bugger.' Years later, when as a student I revisited him, the major and his beautiful, fey wife had been brought so low by family unhappiness as to be scarcely recognisable, though he had not lost his sardonic, sandpaper laugh.

There was a spectacular Falls Pool, the stumbly path up to

it strewn with freshwater mussel shells prised apart by pearl hunters. Even when the river was thin, this was a likely spot for a summer fish, not least because it was permissible to use worm tackle in the great cauldron where they often lay – the preferred technique was to freeline a milk-fed lobworm off your fly rod, feeling for any takes with your thumb and fingertip. (Salmon landed by this method were somewhat coyly entered in the lodge fishing register as having been taken on the 'Gardenia fly'.)

Here I saw my first ever salmon landed, and the fortunate angler was my mother, who, despite a secret and heartfelt disinclination to have anything whatsoever to do with field sports or the killing of animals, had been persuaded to give it a go while the men were fussing over their hip flask, and was rewarded with a 4-pound grilse. I saw Donny Murray cleek it out, and display it, shivering on his gaff. My mother, gamely smiling for her husband's camera, in her navy cable-knit sweater and silk headscarf, had her face duly smeared with clots of gill-blood to mark her first capture, and carefully maintained a pretence of fulfilment and joy, as this was a time when the family was intent on creating an illusion of harmony. She had been a Hollywood star before she married the rising politician, so she certainly knew how to put on a good act, especially as it was not even certain how much longer she might stay with him.

They did remain together, though, for the rest of her life, and that photo went into the family album. But she never again agreed to go out fishing.

*

Except for the sacrosanct salmon parr, nothing much got put back alive.

We had a family expedition one sunshiny weekend to Skerray Bay, where Mr Anderson (who looked like Mr Pastry, but in a roll-collar submariner's sweater) took us out in his lobster boat for a chance at some mackerel. We began by trolling – trailing lures behind the boat – with hickory rods as thick as a hoe handle, their clunky wooden pirns loaded with Cuttyhunk line that held great lifebelt-shaped sinkers and an assortment of feathered hooks that had a rudimentary rubber sand-eel with spinning vane at the end of the rig. To start with, we winched up some squirming coppery pollack, then the motor was cut and we began to drift.

As I dutifully jigged with the handline I had been given, Mr Anderson warned us all, 'Beware the scalders when you're bringing it in.' I peered over the side and saw the morning sea punctuated with red and white jellyfish, like blood-streaked eggs poaching in a pan. If you had a rod you were spared the slimy stings, but when hauling the line by hand onto wooden winding frames you had to watch it. Soon, we were in among a dense shoal of joeys – small, lustrous, strangely muscular school mackerel that came spiralling up from the depths in a phosphorescent carousel and then erupted in zigzag splendour, to be hustled over the gunwales, their velvety scales smudging off as they bounced, thrumming a tattoo on the boards of the boat. If you managed to land one on each of your hooks, the cry went up, 'Full house!' – I imagine a grey-green submarine mansion, somewhere below, with packs of fish bustling between rocky chambers.

The slaughter was exactly what we all wanted. There were four of us children aboard, so it was noisy with the usual rainforest whoops of triumph and bragging. At times, Mr Anderson could not keep up with our demands for his help dealing with the jiggling, enamelled beauties, unravelling the fankles. As he unhooked each fish he thrust an impressive thumb into its mouth and broke its neck, dropping it into one of the baskets where they were rapidly accumulating in a faint blossom of pinkish foam and flared, feathery gills. In any lulls, he set one of the smaller fish on a wooden board and sliced a strip of skin and flank into a 'last' (or 'lask' as it is sometimes called) to impale on our feathery lures, though they hardly seemed to need much enhancing.

His hands bore some stigmata. The palms at the base of his fingers were scored and cicatrised as the result of a youthful accident. He had been out long-lining in his father's boat, and was pulling in and coiling the very end of the tackle when a blue shark appeared and lunged at one of the pollack flailing towards the stern. The shark turned away with its prey still attached to the line, and dived, tearing a dozen or more hooks through the young man's flesh before his father was able to cut him free. I stared at him from beneath the safety of my sou'wester as he calmly told us this, steering back contentedly towards the jetty.

*

Teatime in the dining room was for me one of the highlights of that Tongue stay, as I usually shared it with my paternal granny, the Baroness. While still an Edinburgh lassie she had lost her left eye to diabetes, as a result of which she always wore a daytime hat and veil, and had one of the lenses of her spectacles blacked out. She was sprightly and demure, and had exquisite little hands. Mr White would lay out a proper spread in front of the hissing peat fire: drop scones, rowan jelly and a curious concoction named 'switched egg' – a sweet yellow froth she liked to spoon over toast, the effects of which were presumably countered by the syringes of insulin that she received twice daily for some seventy years. She was cared for by a sulky Italian nurse, and indeed the entire lodge had something of a nursing-home atmosphere to it.

'Will they let me send you cakes, at your new school?' she wondered. At the end of this formative summer, I was going away to board in Berkshire. I replied that I did not know.

'It will be nice and quiet there, in the country, dear.' 'Yes, Granny.' Quite what she thought her eldest son had done to bring the hordes of gawkers and scribblers to the entrance gate of this holiday home, I naturally did not discover; but since she rarely ventured out – except when we trooped *en famille* each Sabbath to the kirk, and sat in the laird's box, where the other faithful were polite enough not to turn around and stare – perhaps she truly felt nothing much was amiss. For me, it was enough peacefully to be in her company, swigging my orangeade while she sipped her Earl Grey.

It must somehow have been decided after my bedtime that

bait-fishing was all very well, but, to be entered properly as a sportsman, I needed to catch something on the artificial fly. Accordingly, I was dispatched to Lochan Dubh with Donny Murray.

The Little Black Loch had an unkempt clinker-built boat pulled way up into the reeds and tethered with a moss-bound painter. Trout fishing was not held in much esteem by the present Tongue tenants, and I suspect the place had not been visited all season. It was a soft morning, the high clouds livid like a bruise, a rising breeze with no cat's-paws of squall. To me the place looked enormous, though it could not have been more than a few acres. In a sloshy rhythm, Mr M. began to bale with a paint tin, its crusty residue the same estate blue as the strakes of our boat. 'Now,' he grunted at last, clambering ashore and kneeling where he had left our mackintosh leggings, the collapsible landing net and a canvas gas-mask bag with Thermos and sandwiches wrapped in greaseproof paper. From his inside pocket he produced a metal fly box, opened it and hissed through his teeth in deliberation.

He flicks a series of bushy patterns so they stand upright in their little clips, telling me the names of each fly – Black Zulu, Invicta, Soldier Palmer. I point to a gaudy peacock one. 'Aye, the Alexandra' – though he does not actually choose it. 'That will do when there's a sun amongst the cloud. You need a broken sky for him, right enough.' He ties on a team of three, patiently showing me the knot, though I can't manage it and am self-conscious about my bitten nails. He hands me the rod – a dark-stained cane 'Marcella', with heavy metal ferrules, and a pronounced, drooping 'set' to it. 'That's us.' He settles in the boat, and the hazelnut water chuckles beneath our keel as he strains against the foot-board and there comes the chalk-on-blackboard screech of oars in their rusty thirl-pins.

Through the thin snake rings I pay out the green silk line behind the boat, the ratchet of my Hydra reel grating. Either side of our wake, in which the flies are trailing, the twin whirlpools from his oar strokes steadily dissolve amid the wavetops. I crouch over the thwart, a loop of line clamped against the pitted cork handle, ready for action. He swings us gently round the central island, then heads towards a bay of weeds. Flinty morning light gleams off the corrugated water.

There comes a buckling of the surface, a bronze flurry and a thrilling pull – all at once. Before I can make a mistake, a dark fish has turned down on the point fly and feels the hook, takes briefly to the air, and jags line away from the path of the boat. 'Aye,' says the gillie, 'rod up, now.' The wooden tip describes an arc of beauty, nodding emphatically as I crank in line. This is no mere burn troutling; when it is lifted from the net and killed, the pocket spring balance sags down at nearly a pound. It had taken a Teal and Green.

'That's a big one, isn't it?'

Donny gives a kindly nod. 'There'll be more yet. We'll maybe get one casting.'

And so we drift broadside to the breeze, and I put into practice some of the overhead casting my uncle had rehearsed on the lodge lawn, an empty bottle of India pale ale tucked into my right armpit to encourage the correct elbow movements. The line whips the water behind me as I strain for even a meagre distance to launch the flies, and Donny hunches for safety, tortoise-necked beneath his fore-and-aft. 'You might slow down,' he reminds me. Occasionally my line makes the right sound in the air, the cast unfurls towards the shore, and the bob fly flirts back on the retrieve, scoring the water. I miss a couple of fleeting rises by pulling it away too quickly, but before the wind dies and the sun comes through we have two more trout, smaller specimens, lying on the boards, thin strips of flaked blue paint sticking to their skin.

With a sigh, the prow slides up into a sandy wedge in the rough grassland where we disembark for a lunch break. Donny flumps down the bag, and hands me my piece. I find a peat 'hag' to sit on, content with banana sandwiches and bottle of bitter lemon. The gillie settles a discreet distance away, unfolds his leather tobacco pouch, and sucks flame towards the sump of his pipe. A little tendril of thin smoke arises.

Stand back with me now for an instant, at the other end of this peat hag while the luncheon break is in progress – back at the edge of things, where so much about fishing seems to take place, particularly in the edgelands of memory. I have been on the water man and boy for nearly sixty years, and they say the child is father to the man: how did I get here from there, exactly?

Largely because those early Highland days have exerted such a lodestone pull upon my life, I now find myself dwelling up a Perthshire glen – a dishevelled grandfather with his scholarly stoop, hunched here over this keyboard. As a travel writer I have trudged up to Everest base camp, and been down in a submarine (there are few depths to which I will not sink, to get a story). I have survived a helicopter crash, a boating accident, cancer surgery and four decades of (blissful) marriage, and have swum my hooks in pursuit of vundu and Dolly Varden, permit, piranha and pike-perch. I may even be growing an adipose fin. Yet I am still as singularly enthralled by the prospect of fish as ever, and that forms so strong a line of enchantment through the passage of years that casting my mind back to Lochan Dubh produces this effect of synchronicity.

For fishing interacts strangely with regular perceptions of time. Firstly, its restorative effects are rejuvenating and seem to bear out that supposedly Babylonian proverb, 'The gods do not deduct from a man's allotted span the time he spends in fishing' (I like this notion, as by that calculation I would still be in my teens), but there is something more, which is a common experience when actually on the water. One can become so transfixed and preoccupied that the everyday goes on hold, and it seems you have stopped the clock. This is why your precious holiday week away appears to be gone in a flash, yet to last for ages. The tyranny of Time itself is challenged, and the realms of the definite and the mundane (seemingly so important in your other life) are temporarily set aside in favour of the thrilling and the unpredictable. Civilians call this escapism, but to me it's more positive than that. It's a minor miracle, wherein time flares out, compresses, becomes elastic. In his trenchant *Poetry in the Making* (1967), Ted Hughes explained: 'All the little nagging impulses that are normally distracting your mind, dissolve ... you enter one of the orders of bliss.' Being transported into some kind of Dreamtime seems to me partially to account for this pastime's famously narcotic effect.

Most things are relative in fishing. Einstein had a nice explanation: 'Sit on a hot plate for a minute, and it seems like an hour. Sit with a pretty girl for an hour, and it seems like a minute – that's Relativity.' Wherever you may be physically, fishing is always there for you in your fancy, like slipping once more through the back of that wardrobe and into a wonderland that operates to a different tempo. So when I consider the great concertina stretch of years now between me and that distant land of the hills around Tongue and a lunchtime picnic, I will settle for the idea that we are linked by some mysterious tunnel in the space–time continuum that is said to connect different universes – and one name for that, aptly enough, is a wormhole.

Not long ago – and mindful in advance of aphorisms about water under the bridge, or never being able to step into the same river twice – I decided to try to find that burn once more (I never did learn its name, if indeed it has one). Not having been anywhere near it in the decades between, it took me a while to locate, and even then it was not immediately recognisable.

Probably to assist some forestry scheme, it had been channelled into a more efficient drainage ditch, and there were no more pots and eddies, sharps and flats, and nowhere so obviously fishy that a boy and his toyshop tackle might begin to cast. I parked my green Land Rover in the new passing place, with its proper tarmac skin and white diamond-shaped sign, and knelt in the heather upstream of where a capacious pipe now took the water smoothly beneath the road. I suppose I might have been expecting something of this sort, but it was still disillusioning. The surrounding moorland looked just the same, however, so there was no doubting the place.

Then, in the run below the bridge, there came a delicate, belly-button rise. On stiffish knees, I rose, too – this time there was no worm rod, but I scrambled up to the car to fetch my 3-weight, casting back once again over the gently disappearing water.

3

OF SMALL FRY

'They neither work nor weep. In their shape is
their reason. For what other purpose except for the
sufficient one of perfect existence can they have
been made?'

Virginia Woolf, an essay on the London Zoo
aquarium, 'The Sun and Fish', from
The Captain's Death Bed (1950)

Although I spend much of my time trying to attract the attention of their heftier cousins, I have never quite graduated from the elementary thrill of catching tiddlers, and I am faintly suspicious of those haughty Corinthians who spurn these building blocks of our sport, or condemn them as a mere nuisance. It is true that small school fish can prove exasperating when they intercept your offering before larger quarry gets to assess it, but on the whole I have retained an affection for what is often called their Lilliputian appeal.

Children seem naturally fascinated by miniature versions of the familiar – puppies, doll's houses, train sets, scale models – literally so, in the case of fish. Being on the diminutive side themselves, they find it reassuring that good things can come in small packages. Writers from Jonathan Swift to Terry Pratchett have played on this alteration of perspective, though I think the formidable author of *Gulliver's Travels* would have greeted with a thin, indignant smile the news that his adult novel became, in many sanitised versions, a children's classic, since he abominated children. Arthur Ransome, that laureate of youthful adventure, and one of my pin-ups as a fishing author, to boot, was similarly conflicted; one river keeper commented, 'He's a grumbly old bugger, and he hates children.'

We might make a marginal distinction between fish that are juvenile stages of species, which grow much bigger, and thus, in the hand, are promises of greater things to come, and bonsai creatures destined forever to be classified as small beer, but which are nonetheless perfect unto themselves. It is to this quality – teleologic beauty – that Virginia Woolf was responding, and which is why they are so often on show, shimmying across their tanks in dentists' waiting rooms, reassuring us that all things will be well, that small can be bright and beautiful, even with the prospect of root canal work. Little fish are Dame Nature's opuscules (let's not drag God into this) and we should not take them for granted.

When the majority of Britain's population was still rural, most children would have been able to recognise varieties of common or garden fish, a traditional affection for which is reflected in their myriad vernacular names. These chiefly belong to what have been designated 'coarse' fish – a caste system that endures as an unfortunate instance of the high Victorian desire for classification, divide and rule, which spuriously based this distinction on supposed table qualities, so that the salmonids ('game' fish) were effectively toffs and workaday chaps such as bream and roach were blue-collar citizens. (I am even less keen on the North American designation of carp and suckers as 'trash fish'.) Some do appear dowdy, but others are resplendent in their own right – the rudd sports ventral fins as red as any street-walker's lipstick, and a perch in the smaller sizes is a fish by Fabergé. The elegant dace was known as a dart, the chub an alderman, the loach a beardie, and the glum tench a shoemaker (I suppose, from its resemblance to a cordwainer's leathery apron). Some were literally run-of-the-mill characters – the unlovely miller's thumb, or Tommy-logge, owes its sobriquet to a squat head that resembles a human thumb depressed from a lifetime spent proving flour. These were like loose change – coppery and silver – picked from the pockets of stream and pond. As their folkloric nicknames attest, they were a natural part of growing up.

Although it is not angling proper, many of these crucial close encounters come courtesy of that trusty implement, the butterfly net. After church but before the egg hunt one Easter at my godfather's farm in Suffolk, the grown-ups, toting their flutes of champagne, were inspecting his woodland plantings (a shy, velvety-voiced parliamentary colleague of my father's, he was prominent in the Royal Horticultural Society), and I, as the only child present that weekend, was allowed to make my way down the bank below the terrace, avoiding the exotic, rhubarby leaves of the gunnera, the gas-jet glow of crocus and the daffodils swaying like a gospel choir, to prospect idly in the formal, oblong lily pond. There, along with much fine, hairy weed, I scooped up a pink-breasted stickleback and decanted him into the proverbial jam jar. This was much better than any tin bucket full of ovoid chocolate. I felt as if the entire universe were peering over my four-year-old shoulder at the puckish, moon-eyed captive viewed through the lens of his temporary prison. I have perhaps never lost that fundamental, childish desire to *possess*.

In his 1837 novel *The Posthumous Papers of the Pickwick Club*, Charles Dickens records that his protagonist submitted a learned paper containing, 'Some Observations on the Theory of Tittlebats', but, disappointingly for scholars of the Gasterosteidae, the author furnished no further details. This bristling, surprisingly belligerent goblin of a fish is our paradigmatic tiddler – indeed, the term derives from a nursery version of tittlebat, his many other names including sharplin, banstickle and Jack Bannock. Some of the three-spined variety are tolerant of saltwater, even anadromous, and their flanks are armoured with bony scutes, while dorsal and ventral fins are daggerish with erectile spines. Tom Tiddler seems always alert, incessantly strumming his pectorals like some hummingbird on patrol.

In breeding season, the male puts on the Ritz, donning an epigamic courtship dress which comprises a throat of glowing crimson and boldly aquamarine irises. He looks as flash as a rat with a gold tooth. Unusually, it is the male who constructs the nest, gluing together plant detritus with a mucus called spiggin that is secreted from his kidneys. In defence of his marital bower he can be quite ferocious, and in this mating garb nicknamed a 'robin' – aptly so, since his avian counterpart (which, with its misleading associations of Yuletide benevolence, is really no more than what birders call a Little Brown Job, but with ace PR) also displays the kind of deceptive viciousness worthy of a game of vicarage croquet.

My occasional pursuit of the tittlebat has continued into later years. Since 1974, I have spent happy summer months on the Isle of Harris, where my wife's family has a croft on the stony shores of the Minch, miles from the nearest road; and here, with or without offspring, I prospect the craggy shoreline pools and intertidal zones with my favourite brand of butterfly net – an agreeably robust one I found in an Ullapool sweetie shop, the simulated bamboo shaft moulded from bubblegum-pink plastic. Among the bladderwrack and sea lace we actually harvest small fry, mackerel and cod (the house record stands at thirty-three in one session, plus a bonus sea hare), along with Mekon-browed blennies and shannies, creamily badged butterfish, and assorted smidgeonettes that are borrowed from Neptune, grandparentally admired, then returned to the salt. At the turn of the tide, if the sea breeze should drop, crystalline shrimp can be individually targeted as they tiptoe over the kelp; these are kept for the pan.

A burn runs down beside the house, and its crannies harbour elvers, which can be guddled out from between the cold stones (one year I imported an aquascope from Florida, to assist in peering underwater, but this was deemed to be cheating), along with estuarial three-spined stickles, which, during times of spate, ascend to the Loch of the Rowans up the brae above, where there is also a resident population of the ten-spined variety (Britain's smallest fish), upon which the brownies batten, waxing fat and elusive. One morning, long after our three children had outgrown their enthusiasm for this type of watery safari, I was dredging around the margins of the loch, kitted out in chest waders, tropical Polaroids and a billfish cap, when I looked up to see a windswept blonde woman – quite an unusual sight around this remote headland – regarding me with some astonishment. I later learned she was an Aberdonian academic engaged in a study of migrating mink movements, but quite what she thought this middle-aged man was doing paddling alone with his pink net, I never did discover, as, with a fleeting wave of acknowledgement, she made for the hill as rapidly as seemed polite. There had been no time to explain the spiggin to her.

The pursuit of such trivial quarry is never going to be the stuff of saloon-bar rumour or the zoology of the fabulous – they are mere foot soldiers alongside the lordly salmon or warrior marlin – but sometimes we ignore at our peril the qualities of being obliging and abundant. Such small fry can also save the day – they become silver linings. Many a canal matchman has been grateful for small mercies when it's come around to weigh-in time (being temperamentally disinclined, I hardly ever fish in competitions, but I won a recent Kennet tournament with a brace of gudgeon, neither of which measured 5 inches long). Judges sometimes have to work to minute tolerances: the total bag of one Lancashire champion, back in 1878, 'turned the scales at a fraction over a quarter of an ounce, plus two rabbit shot', though it was enough to win him a new teapot.

There are still occasions when catching a mass of small fish gives me pleasure and, though in general a preoccupation with arithmetic is not conducive to sporting happiness, the opportunity to amass sheer numbers is sometimes irresistible. I have succumbed to this most often on the bonefish flats, where a freak convergence of 'schoolie' chicken bones has allowed one to rack up over forty before heading elsewhere. Once, on a thoroughly overpopulated wild trout loch (near a lodge now lost to fire) and under instructions to thin them out, we called a halt to our angling in mid-afternoon when continuous success was becoming monotonous and it was starting to feel like 'Groundhog Day'. Our boat had made a basket of 101 little brownies ('eldrins', as they used to be called), just to see if it could be done. The lodge hogs ate most of them, but it must be admitted that – even though I don't much like eating fish – the flesh of such 'breakfast' specimens is sweet, especially if dipped in milk, rolled in seasoned oatmeal, and fried, al fresco, in bacon fat. I believe there is an argument that you shouldn't be fishing in the long run unless you intend to keep and eat at least some of what you catch.

In the bad old days, few anglers thought twice about the taking of such biblical numbers – such is the tragedy of the commons. Dr Knox, the nineteenth-century Edinburgh anatomist, was once so convinced he had caught the last remaining trout from a favourite burn that he took the unusual step of returning it. Burke and Hare, who supplied him with cadavers, would probably not have understood this rare instance of compassionate catch and release.

Ethics and taste can shift abruptly, and for centuries before the Victorian subdivision of freshwater fish into distinct classes of sporting desirability, practically everything was dished up with relish, as it frequently still is in continental Europe (though most sensible folks have always drawn the line at chub). Now rather maligned, the gudgeon is a good example. This olive-sheened, blotchy little fellow is a bottom feeder with an agreeably catholic appetite, sporting twin lip wattles and a cutely crescent mouth. He may be no great beauty, but he was once a notable favourite with Regency parties on the Thames, where young ladies and their admiring swains were permitted to angle unchaperoned in punts moored to ryepecks by professional fishermen, who pre-baited the swims and periodically raked up the gravel to keep the shoals of gudgeon on the feed – a practice known as 'scratching their backs'. So multitudinous were these fish (before the coming of the railways allowed access to the open countryside for all and sundry) that even the least-accomplished groups might account for thirty dozen or so in an outing, and these were later served up with champagne as part of a bankside feast, a small-fry fry-up – goujons was a word and a delicacy introduced by our Norman conquerors.

These voguish occasions lent themselves to symbolic illustration by artists showing young men being hooked by feminine wiles, and this modest little fish became a byword for gullibility and the easily duped. I reckon this gave Shakespeare the clue for naming Shylock's servant-clown Launcelot Gobbo (*gobione* is the fish's Italian name, and the play is set in Venice), and he must have had these teeming shoals in mind when, in *Hamlet*, he coined the term 'groundling' to describe the coarser members of the Jacobean theatrical audience, who milled around the pit at the foot of the stage, snappers-up of the playwright's unconsidered trifles – *Grundling* being the German word for gudgeon. Even the humblest extras have their own history.

Going after what some Americans term 'popcorn' fish can offer you a refresher course and rekindle your *joie de vivre*. You may not need the gladiatorial sophistication of the gear required when taking on glamour species such as steelhead, permit or 'cow' tarpon, but there are advantages in recalibrating your kit and getting back to basics, instead of sallying forth as encumbered with gizmos and caboodle as a high-rise scaffolder.

You frequently use a float – that quintessence of our pastime, beloved by cartoonists as the emblem of concentrated fatuity masquerading as patience, a truly ancient device that may be as simple as a single straw, a tuft of wool from some fence wire or a discarded wine cork. There have been outlandish variations that incorporate glow-worms or fish-finding detectors, and the spectrum of ingenuity is impressive; but few of angling's artefacts offer so much aesthetic pleasure as a trim, hand-rolled telltale – that basic float fashioned from quill or balsa, the making of which remains a cottage industry. Telltale was also the name mariners gave to strips of fabric hung about the stays to indicate changes in breeze, and nicely evokes the mesmeric experience of monitoring the nuances of its behaviour on the surface as you watch for the least quiver, uplift or curtsy – was that shudder a passing zephyr, or the brush of a tench's lips? Intent upon the bright ember of your float tip glowing against the dark skin of the water, it acts like a wire-tap into that other world – 'pleasing in appearance, and even more pleasing in disappearance', as H. T. Sheringham peerlessly expressed it (*An Angler's Hours*, 1905).

Quite different from my first Highland burn experience was a visit the following summer to Dedham Mill, where we float-fished for numerous perchlings, using the cockspur worm and a basic adjustable bung. These were but entry-level, pipsqueak specimens weighing a few ounces, but their pugnacious vigour as they stabbed the bait away with their carpet-bag mouths, and their ornamental quality (part cryptic coloration, part Carnaby Street swagger) as we removed them reverently from the knotted orange meshes of our crudish keep-net were enough to make them seem – in the phrase of Sam Goldwyn (originally Goldfish) – 'colossal in a small way'. Back in London, I persuaded my mother to take me to the Serpentine where you could then still fish below the road bridge on the southern shore, and there I reeled in plenty more ravenous little hog-backs (the loyal Mrs Bivar kept my maggots – 'gentles', as they are known piscatorially – in her fridge, sustaining them with chopped liver).

My apprenticeship was expanded by two childhood holidays we spent aboard another uncle's converted motor patrol vessel, *Seafin*, which was based in the Mediterranean. I was allowed to fish off the harbour walls just after dawn, with a bamboo pole, a length of nylon attached directly to its tip, and some mealworms stored in newspaper. I now realise that what I was shaking off into my pail were mostly goldline (*Sarpa salpa*), sometimes known as dream fish, as eating them can give you acid-style hallucinations. They were not welcome in the galley, so I offered my modest catches to various stubbly, nicotined old men in grey caps who were apparently fishing for the pot.

The greatest privilege, it seemed to me, of being sent away to board at Eton at the age of twelve was that the college had exclusive fishing rights to a great stretch of the left bank of the Thames, below Windsor. Though never himself an Etonian – unlike James Bond, Captain Hook (prosthetically part-angul), a peculiarly large number of prime ministers and (if you believe certain of the more old-school alumni) God Almighty Himself – Izaak Walton liked it here, and angled at Black Potts as the guest of his mentor Sir Henry Wotton, a retired diplomat who was then Provost and who gave him the idea for a little manual that was to become one of the most reprinted books in the entire history of British publishing. Being a butterfingers on the pitch, and a gangly, physical coward on those playing fields where the Battle of Waterloo was supposedly won, I was never much in demand for team games, my athletic contribution being limited to bringing on the half-time oranges. Acne had invested my sullen teenaged face with an epidermis like recent roadkill, and during my leisure hours I opted to prowl around the willow runs and backwaters, rods in hand, intent on freelining cheesepaste for chub, or livebaiting out of season for pike (we claimed, when challenged by the college policeman for violating the 'fence months' – the close season when fish are breeding, and protected – to be targeting Thames trout, for which there was even a school cup, last awarded shortly after the Great War). For some reason – probably dating back to a founder's proviso in the fifteenth century – our college permits included the stipulation, 'Carp may not be taken.'

Our preferred bait was the bleak. Its name derives from the Old English *blaece*, which has nothing to do with being drear or dismal, but means gleaming white – as in bleach, or the bleak midwinter. Father Walton himself certainly appreciated them, and his description is one of the most graphic in all his *Compleat Angler*: '... a Fish that is ever in motion, and therefore called by some the River-Swallow ... his back is of a pleasant sad or Sea-water green, his belly white and shining as the Mountain-snow'. Indeed, with its streamlined, gracile body, steely greenish back and iris of pale gold, there is nothing coarse about the bleak – it is almost like an Elzevir edition of the mighty tarpon. Once so abundant in Medway and Ouse that they were netted out as fertiliser, this bright little flip of a fish cavorts on the surface after insects with such verve and abandon that it was known as 'mad bleak', and in the bleak midsummer we used to 'whip' for them, Walton-style, with lightweight fly rods and a size 18 Black Gnat. Children have long enjoyed chasing the river swallow: Ausonius records they were 'prized booty for schoolboys' hooks' on the Moselle river, back in the fourth century.

Walton added that the 'blay' was as tasty as a house martin (which were also then angled for by casting artificial flies off church steeples), and though I have tried them fried I found them disagreeably bitter. Like many fish, their flesh varies according to where they have been feeding, and *Alburnus alburnus* has an occasional penchant for the outflow of sewage drains. In Victorian times they were sold, salted, as 'mock anchovies', and were a breakfast favourite of one of my heroes, that enterprising all-rounder Harry Cholmondeley-Pennell, poet, crack shot and deviser of the Pennell fly series; I have one of his library volumes, and from the photographic bookplate he stares out forbiddingly, as if you had recently importuned his sister.

A humble acrobat on the sporting stage he may be (perhaps

a diversionary act during the interval) but the bleak was once of considerable economic significance in the artificial pearl industry. Struck by the ease with which nacreous scales were dislodged from a haul of *ablettes* culled from the Seine and being rinsed for the table, a Monsieur Jacquin rediscovered in 1656 a Chinese process for grinding down bleak scales with ammonia into a precipitate that was used to create gleaming moulded gypsum fakes, though when these were found to melt under the heat of fashionable chandeliers he invented a method of coating his paste onto the inside of hollow glass spheres (the guanine crystals impart that silver lining). At its peak, his manufactory was producing ten thousand such scintillating counterfeits each week, but as it needed seventeen thousand bleak to produce a single pound of this squamous *essence d'orient*, eventually the scales were replaced with powder from the swim-bladder of the smelt-like Argentina pearl fish that inhabited the Tiber – though it is said some Venetian glassmakers today still rely on the original *essence* to imbue their handiwork with a patina of antiquity.

Some of these pearl factories may have been referred to as Bleak Houses, but – though he does allude to 'silver linings' – there is no scaly connection with Dickens's novel, its title being the address of his lawyer Jarndyce, signifying the pervasive gloom of those hard times.

In his diary entry for 16 June 1872, the Very Revd Patrick Murray Smythe recorded how, at the age of eight, 'Coming back from Church, caught a minnow in my hands – I ate minnow.' I doubt there is a single sentence in the entire *biblio-theca piscatoria* that so deliciously encapsulates the spirit of our sport (the diary runs to eighteen published volumes).

This near ubiquitous, minimal member of the senatorial carp clan is truly handsome, dashingly tailored in silver and green glossed with gold, as debonair as any uniformed foot-man – yet a byword for the least among equals. It has a dainty, toothless, peppercorn mouth, and is a gentle, inquisitive nib-bler, who manages to appear both delicate and muscular, with a satin feel to the tiny-scaled flanks. In April, as the spawning season approaches, the male minim develops a sooty 'chin-strap' and a pink belly (he is also known as a penk) and grows a rash of curious white tubercules around his head which are believed to attract females – the absolute reverse of acne. The minnow boasts a nifty Linnaean tag, *Phoxinus phoxinus*, that denotes his tapering shape (folk names include baggie, shadbrind and Jack-barrel, which sound more like a gaggle of Falstaffian low-lifers). His is arguably the archetypal walk-on part in the nation's aquatic drama.

'Little things should not be despised,' observed Nicholas Cox in his *Gentleman's Recreation* (1674), before going on to give a recipe for a minnow 'tansie' – a sort of Eastertide frittata flavoured with cowslips. Before supplies of fresh sea fish were readily available away from our coasts, a dish of minnows was often welcome as a form of freshwater whitebait, and records of several mediaeval banquets list them being served up by the gallon. The egregious Victorian naturalist Frank Buckland supplemented his diet when a scholar at Winchester with pickled chalkstream penks, though the plate of *phoxinus frites* I once sampled was barely palatable; it is said you need first to remove the gall bladder, but somehow life seems too short to peel a minnow. However, alive or dead it makes a bait nonpareil for chub, perch and trout, and many artificial lures, fashioned from silver, glass or quill, have been designed to mimic them. Minnows will brighten attractively if kept awhile in a palely painted pail (what Ransome dubbed a 'whited sepulchre', and which I cannot forbear calling 'A Whiter Shade of Pail'.) One Regency 'squarson' used to store his livebaits in the church font.

We moved to a ramshackle Hertfordshire farm when I was nine, and through its pastures ran an unkempt section of chalk stream which appeared not to have been fished in ages. It could not have been a further cry from the barbered beats of Hampshire, with their splendidly appointed luncheon huts, and footbridges neatly upholstered with chicken-wire, nor was it one of those fluvial fly factories with freshly hatched duns drifting down as if on a conveyor belt. There was certainly no river keeper to oversee proceedings, which was just as well since I had the run of the place unsupervised, and hunted down by various methods (none of which included presenting an upstream dry fly while crouching on a leather kneeling-pad, dressed in a Norfolk tweed jacket like some stereotypical Edwardian) most of the long, slightly esoxian-looking brown trout that inhabited holts and boltholes amid the bramble bushes, logjams and unpollarded willows.

One spring, on a stickly set of shallows, I spotted a school of minnows and I set about sight-fishing for them with a scrap of worm on a size 22 gilt hook. In the clear water, I could plainly watch the bait being gobbled. I hoiked out a dozen, slew them and laid them in two empty air-rifle pellet tins. I had been reading up about the use of the baggie on drop-tackle in the Scottish Borders, and adapted it for this smaller southern stream by cutting off the tail section, nipping a split shot inside and threading a small treble through the vent; this rig was lowered into likely nooks off my fibreglass fly rod, and resulted in some violent encounters with brownies, the largest of which weighed 3 pounds and 12 ounces. These days, I sometimes yearn sacrilegiously for my old minnowtail tin, its interior anointed enticingly with pilchard oil, when I am loitering, unobserved, by some hatch-pool on the pedicured banks of the hallowed Test.

In my O level biology year at school, I came up with the idea of trapping minnows from the Thames and transport-ing them for use on the few remaining trout back home. This was easily enough done, using an empty wine bottle with the bottom knocked out, gauze fastened over the neck, and a piece of red wool fluttering with the current inside. The dozens so garnered were killed and tipped into a series of screw-top plastic storage jars which I topped up with some home-made formalin mixture. My parents had just taken delivery of a carmine-coloured Austin Maxi (we were turning into a racy lot, back in those Swinging Seventies) and when my mother came to collect me and my belongings for the first day of the summer holidays, the intense July heat acting on my amateur miscalculation of chemical mix in the preservative fluid generated an eye-watering explosion from one of the containers as we were driving through Totteridge, with the result that the interior of their vehicle was never again quite in showroom condition, nor was my domestic reputation as a teenaged ichthyologist.

The cult of piscatorial minimalism probably arose around the eighth century A D in feudal Japan, when leisured but cerebral nobles fished for *tanago* – a minuscule, bitterling-like creature with a pinkish-blue sheen – from specially constructed waterside temples (*turidomo*) using lines made from a single human hair. Some Zen anglers further refined this experience by using no bait, reasoning that the actual hooking of a fish might impair the serious experience of contemplation. Modernised versions of such practices have become recently voguish in certain North American circles, where microfishers pursue a spectrum of palm-sized freshwater species with slender bamboo rods and microscopically sharpened hooks, the goal being to catch the tiniest possible fish (ideally your quarry would be smaller than a one-yen coin). This should not be confused with the dubious fetish for landing large specimens on the lightest possible line, which is all the rage among those intent on establishing world records, nor is it identical with *tenkara* techniques, which have also proved popular in the West. The latter, which developed in the hill country of Japan and involves a line attached directly to the pliant rod tip, with the use of a simple selection of reverse-hackle wet flies stroked ingeniously across the current, is an intriguing method that requires a profound familiarity with how to read the water and was designed for peasant households to catch the elusive little *amago* and *iwana* from bouldery streams, and may be intriguingly adapted for our moorland troutlings, or lowland school grayling. The attendant rituals can be complex, and include the bankside drinking of sake infused with fishbones, but these trends challenge the tyranny of avoirdupois and keep us mindful that refinement and finesse are central to the mindset of the pursuit.

When our eldest son, James, was seven we went on holiday to New Hampshire. Our neighbour Smitty lent us his canoe, and we set off in search of panfish. Paddling harmoniously along the lake shore, where creeks came ambling in, we flip-cast a twirling blue Mepps lure towards the reeds, and picked off a few sunstruck perch. This was a plainsong version of the sporting experience, one that murmured its pleasures – simple and delicate and memorable. Back at the house, a ritual barbecue was prepared for my son's first ever fish.

We ate perch.

4

The Airman and the Admiral

'My life would go on like that forever, in a world
whose mysteries could never be solved.'

Henry Williamson, 'The Boy Who Loved Fishing',
from *A Clear Water Stream* (1958)

Our annual summer pilgrimage north to my uncle's estate
in Sutherland began at the Olympia train terminus in
Kensington, where the cars were loaded onto the overnight
piggyback service to Inverness. Glen the retriever would be
delivered early to the guard's van, then the party went for the
first sitting in the buffet car – waiters in short white jackets,
their thumbs in the tureen, would serve up minestrone soup
followed by mutton, and, aged nine, I was allowed to collect
the empty liqueur miniatures before we returned to the

compartment, with its fragrant po-hutch (housing the chamber pot) and round suede wall patch to harbour a fob watch. Next morning, wrenched from your locomotive dreams by the gaoler-like rap of the cabin attendant's master key against the cubicle door, you were in the Highlands.

The Lodge was a Victorian shooting box of minimal architectural distinction, belonging perhaps to the psychiatric baronial school, its heavy, irregular exterior set off by monkey puzzle trees, a mossy lawn and an embattled garden. The household was run by Mr White, who was effectively the *genius loci*: neat, ubiquitous, kindly and redolent of Vitalis, he was a master of innuendo, immaculate, handsome and loyal, with an abiding fondness for the twin Grant brothers who owned the village store. He had two friends – also both butlers and, like him, called George – who took their holidays to join him at the Lodge, so the hill picnics were always colourful. At the time, I did not quite appreciate what a period piece we made, but I rather suspected not every boy spent his summer like this.

On arrival, my first thrill was to unload the Vauxhall shooting brake and arrange my outdoor equipment on the shelf assigned to me in the gun room – a place spectacularly caparisoned with sporting apparatus, including a glass-fronted cabinet housing Purdeys (my uncle was a director of the firm), horizontal racks of Palakona cane rods rigged and ready, wooden winders for drying silk lines, a cupboard containing a squadron of pre-war Hardy Perfects with serried ranks of oxblood Neroda fly boxes, and a framed photograph of Uncle Harold with a vast, 185-pound tarpon, which he had killed trolling off the coast of Sierra Leone. There was also a brace of cartridge bags fashioned from the skin of a rogue lion he had been obliged to shoot when he was resident minister in West Africa in 1944.

Presiding over this area of the house was the keeper Mr Murray (yet another George, like Mr White, though our family never called him by his Christian name, either). He had a large, softly folded face with a lower jaw slightly offset, as if he were permanently deliberating something, which contributed to his zigzag smile. His hair, once rusty like a hill fox, was now the faded hue of preserved ginger. From the start, imagining myself in some way his sorcerer's apprentice, I sought this quiet man's approval, his conspiratorial chuckle scarred by years of the Silk Cut he frequently cupped in one hand. He taught me how to spot grouse peeking from the heather as we drove the moorland tracks, and from him I learned the trick of lubricating the metal of your male ferrule by twirling it in the wing of your nose before putting together your rod. He and his slow, melancholy wife, the housekeeper, had no children, and it's possible he enjoyed my absurd eagerness. George Murray was certainly my first mentor.

The estate was administered along traditional Edwardian lines: His Lordship received in person the annual crofting rents, taking a room at the local inn, and the chief attraction of those thousands of acres was the pursuit of modest numbers of grouse over pointers, which took place three days each week, the others (excepting the Sabbath) being spent fishing. On the Glorious Twelfth, a charger of roasted birds would be paraded around the dining room by Mr MacDonald, the retired pipe major. Evenings were always semi-formal, with the men in smoking jackets and suede slippers, our meal followed by bridge or dumb crambo. My uncle was a committed reader (and also the author of several volumes of memoir) but my father, being dyslexic and chronically restless around books – his perennial phrase of literary criticism was, 'Ah, the Romance of Words!' – had a taste for more extrovert entertainment. He preferred to organise singing sessions with Jack Buchanan records, or amateur conjuring nights, where he was in his element, something I seldom saw at home. He was as thick as thieves with his sister, who encouraged his party behaviour, doubtless because my mother seemed to disapprove of it (though outwardly civil, these two formidable women maintained a brittle truce). As the youngest in the household, and an only child – a 'one-fish', as they say in pidgin English – I looked on all this with some wonderment.

On Wednesdays, when the staff had a night off, we packed into the blue Bedford minibus and drove to a nearby hotel – a place of velveteen gentility, whose proprietor was a moustachioed major from Central Casting, in a tweed suit, and whose wife, according to Lodge gossip, was endowed with three bosoms. Starters would comprise tomato juice, or a dish of melon balls dusted with ginger, and there was habitually Mateus Rosé to drink. My uncle, though a connoisseur of champagne, had no taste for still wines whatsoever, and his claret bins were drearily stocked, but my father – who would anyway foot the bill – used to pretend to savour the very ordinary pink: 'I call that pretty good, Harold.' 'Cheer-o,' my uncle would reply, as they used to say in his mess during the Great War.

He had no grandson, and I no grandfather: I did not have to search far for my hero figure (I did not then appreciate my father's sterling qualities, as is often the way). Uncle Harold appeared to me to be not of the common clay. He had a formidable aspect, a bulldog look with a large philtrum and pronounced marionette lines etched down his jowls, reflecting an occasionally peppery disposition. In 1905, he had been expelled from his boarding school for biting the mathematics master. Even his deafness seemed impressive. In his early aviator days, they had not yet developed the interrupter gear that allowed pilots to fire a machine-gun through the propeller, so they were obliged to shoot sideways at the passing Hun with a pistol unholstered from a Sam Browne. When he joined the Royal Flying Corps the life expectancy of a flying officer was three weeks, but fortunately he was a good shot, with eleven confirmed kills. He would happily aver that he would rather be deaf on a grouse moor than hear a pin drop in a bath chair. He seemed to me a man who, under any circumstances, would send sparks scattering through the darkness.

One of his prized possessions was a mighty hip flask given to him by his Victorian godfather. With a capacity of three-quarters of a bottle of spirits, it brought a keen light to the eye of many a gillie over the decades. This notable vessel was named the Admiral, and its sobriquet derived from an incident in 1941, when, as Undersecretary of State for Air, he was returning from a conference in Moscow (where he had enjoyed cocktails with Molotov) aboard a cruiser from Archangel that encountered heavy seas. The captain repeatedly went below decks 'to consult the Admiral', but it transpired he was merely swigging grog to boost his morale. 'Consulting the Admiral' became a familiar bankside ritual, though my uncle was a stickler for standards and did not invite me to take a drink from its silver slip cup until I had turned twenty-one – by which time my alcoholic horse had already bolted, and its stable been converted into a wine bar.

On the back drive, just by a ruined kennel block and not far from the bothy where once I heard spectral footsteps, there was a single, wild gooseberry bush. In late August, when the hills had turned a smoky mauve, my mother and I would visit it and scoff the exquisite little red fruits straight off the branches – a taste I still associate with the tristesse of summer's end. Throughout the rest of the year I would yearn for those precious, intoxicating few weeks in a well-ordered, remote refuge where I felt so replenished and reassured. I longed for it to last for ever, and I suppose I have been seeking somewhere similar ever since.

At my dismal Berkshire prep school, every Sunday after chapel we trooped into a converted army hut to engage in compulsory letter writing home – the pages vetted by the master on duty, in case of any covert pleas for rescue. My mother kept many of these, one concluding, 'I just want to go fishing' – an appeal perhaps against the cross-country runs, Spam fritters and quasi-Spartan absence of comfort (not genuine hardships, but certainly not appearing privileged at the time) which would gradually educate me into the sort of chap who could become a colonial district officer, or supervise rubber plantations from beneath his solar topee. Despite family precedent, I never dared bite any of the teaching staff.

My summer apprenticeships at the Lodge began with trouting on the burns and in a hill loch, though Uncle H. was essentially a salmon man and had little time for brownies. On my ninth birthday I had acquired a basic cane rod and rather a fine Hardy fly reel, and I was more than content with my scale of operations. One Easter holiday, however, we were lent my uncle's beat on the Aberdeenshire Deveron at Ardmiddle, staying at a hotel that seemed disconcertingly exotic – its cocktail bar was a converted dungeon and the menu featured elaborate curried dishes, since its proprietor had once been major-domo to the exquisite Maharaja of Bundi.

My father was happy enough to join in these family parties, but he was never quite passionate about fishing. That March there were still kelts about, of which he caught several by spinning, showing me their nightmarish cargo of gill maggots, though he also landed and promptly killed one beautiful springer ('fresh-run, blue-green and silver-bright – of all shapes surely the most perfect in Creation', as the novelist Neil Gunn describes it).

Just above Ruddiman's Rocks, I had my first Close Encounter. Left to my own devices – unthinkable for any

ten-year-old now – I mooched my way down a slow run, fling-
ing out a silver Mepps, a metal lure which had been attached
to my borrowed Milbro glass spinning rod, along with a green
spiral lead, bent like a banana. My mother was upstream
in the hut, with her Jean Plaidy paperback, and a packet of
filtertip Gitanes. My idle head full of the Monkees hit I had
heard on the car radio, I spun my way mechanically towards
the little groyne, above which there was a steeply shelving
pot, and as my lure twiddled its way up from the depths
towards my boots a shape struck swiftly from the murk and
streaked away with my metalwork in its mouth. I reared back
on the little rod with the vigour of some Montauk surf fisherman hooksetting a striper, the Zebco clutch squeaked like
a stricken rabbit, a sizeable fish rolled on the surface, and my
line parted at the knot below the swivel. It may have been a
well-mended kelt, and anyway my subsequent account lacked
finesse and was met with a mixture of indifference and incredulity. But from that moment, I could say, 'I'm a believer.' That
thrill of sudden attachment was to run through my boyhood
like a bright thread.

To improve my blood-knot tying, Uncle H. later presented
me with a wooden yellow belly Devon minnow, one of its fins
broken off but complete with wire mount and treble hook, the
idea being that I should practise my technique with a length
of coarse monofilament that I also dutifully kept in the basket
beneath my dormitory bed, until Pauline the matron confis-
cated it all as constituting a danger to small boys.

I was just in my teens when an August spate swelled the

little Fleet river, and I was able to fish it for the first time. That usually level-headed High Victorian pundit Francis Francis reckoned in 1874 that 'the Fleet is a small salmon river not much worth notice', but it is one of those waters that will surprise you if you live on the spot and can cover its otherwise shrunken pools within a few hours of a freshet peaking – as it had that morning. Under muscle-bound clouds and with the scrubby field gasping with saturation underfoot, Mr Murray and I made our steady way towards the Stepping Stones.

The river was brustling, brimming over its banks, and there was a submerged barbed-wire fence between us and the main current, which swerved in impressive crenellations around the eponymous, sunken stones – *Stairean*, in the Gaelic, although the name of the river itself has an older derivation, *Fleot*, signifying an inlet from the sea. The water was the colour of bitter ale. 'She's falling, mind,' murmured my mentor approvingly. I had been kitted out with a 12-foot spliced greenheart Grant Vibration rod, with a slow, heavy, nodding action, but in the event a line of no more than 10 yards was required to reach the lies being pointed out to me, which was just as well since my casting movements resembled throwing the hammer rather than a matador's cape work. Ten minutes into the pool, there came an unmistakable 'rugg' to my line, and finally I was fast into a 'fish'.

For more than an hour a sullen tussle ensues. I have of course no experience of fighting fish, have never heard of the 'minute per pound' formula for playing salmon, nor how to walk one upstream or apply side strain, or keep his head up.

'You're doing grand, Mister David,' says Mr Murray, but the blood is beginning to thrum in the cave of my head, and as the postie stops on the sheep bridge and waves his encouragement I feel the situation slipping away from me (and not for the last time in my sporting life). 'I think I'm going to lose it.'

We still have not seen the fish when Uncle H. saunters up from his lower pool, smoking. 'Decent fish, Murray?'

'Been on a while, m'lord.'

'Giving him a bit of fun, anyway.' He unslings the gaff from the gillie's back, removes the champagne cork and inspects the point. The fish continues to plane against the thick current. Then something wallows, in a squirl of water – and there's nothing wrong with the old aviator's aim, as he leans out over the drowned wire, and in one fluid movement cleeks the fish onto the grass, and he is ours.

At 9 pounds 2 ounces, he was indeed a respectable salmon for that stream, a cock fish carrying just a blush of colour, the single hook of our claret hackled Dee wing fly neatly in his scissors (the hinge at the corner of its jaw). How did I feel? Given that until then my experience of triumph had been limited to success in Latin construe tests, I felt as heroic as if I had single-handedly liberated some wartime village. I was ecstatic – standing outside my own body, as if my blood had been carbonated, in that altered state I have enjoyed so often since. I had excelled. I belonged. Back at the Lodge, the clack and babble of its borborygmic plumbing led to the deep bliss of my piping-hot bath, the water sour with peat stain (or was it really pipe rust?), and in my flared cords and crew-necked jumper I descended to a hero's reception. I was presented with the rod, its original leather bindings, and the venerable cloth bag – I have it still, but have not been tempted ever again to test its peculiar magic. Nor, as it happens, did I ever get to fish that river a second time, but maybe I didn't need to. I might have failed to be awarded any school colours, but it seemed that on one field, at least, I had won my spurs.

In these formative years, the rivers I fished, just as the books I read, influenced me for good; so it is that the dramatic and challenging Shin remains my mental yardstick whenever I consider the archetypal idea of a salmon river. With its bouldery gorges, precipitous pots and moiling runs, there is something primaeval about the energy of the Shin – even its name derives from the pre-Celtic *sindhu*, simply 'river', the same stem for the Irish Shannon. In the lower reaches, with no gravel to absorb any of the current's energy, the bedrock receives the full, raw force of its flow as this ancient watercourse hacks and hurtles its way down from a vast loch, harnessed for hydro power but regularly releasing compensation water, so the beats are nearly always fishable even if that artificial top-up may be stale and under-oxygenated. Its celebrated focal point (the centre of where my uncle rented his sport) is the baleful Falls Pool, a churning hellbroth that holds up running fish which flash like spearlight as they strive to leap, and then swim up, a 12-foot flume through a gullet of dark rock, twisting and bucking against the torrent like that Irish hero Finn McCool with his legendary martial arts manoeuvre, the 'salmon leap'. It is said divers have discovered more than one human skeleton down there, twindling and circling upright, caught in the cycle of undertow.

I was occasionally allowed to follow Uncle H. as he fished down a pool, casting behind him, so I had abundant time to observe how he addressed the waterscape, which sometimes involved grasping a wire with one hand and hanging off a rock to deliver your line. The resident gillie was first Willie Macdonald (a great advocate of the Green Highlander fly), then his successor Hughie MacIntosh, a pawky and shrewd companion, who believed in laying siege to individual fish, and teasing them with a Collie Dog. This was naturally all new to me, and it proved quite a hard school in which to learn, though one of the valuable lessons was the need for precision and concentration, and the benefits of casting as well as you can. The Shin is renowned for its race of large salmon, and although they tend to be quite showy they were by no means free takers. Sometimes we would spend half an hour ringing the changes in a single run that looked promising, crouching or standing well back against the rockface. These early combat tactics have since stood me in good stead elsewhere, on quite dissimilar rivers.

So many salmon anglers forget to stop and look before approaching a pool – eager to get in at the neck and lengthen their line. First, Hughie stressed, you should squat down and survey the scene, take in the contours of the pool – I now know this process as 'reading the water', and it is an attempt to attune yourself to the rhythm of the stream (though it applies equally to standing waters, which are rarely still). What is this particular script telling you, as it scrolls on by, articulated by pleats and glides, punctuated perhaps by the odd fin, or other anomaly in the flow? You do not read it as some instruction manual, but more like a detective saga containing clues (some of which may be misleading) – enigmatic, encoded, intriguing. You may need to feel your way below the surface narrative, too, especially if using bait, exploring by Braille-like touch, for innuendos and subtle variations in the plotlines. The angler looks at the water in a particular, somewhat strabismic fashion, but his attention to the *roman-fleuve* can be deeply rewarding, whether or not the finale delivers a fish – an assertion that seems a load of fenks to the civilian. I believe this applies to all types of angling; I have attempted to keep up with my reading.

There was in those days a ramshackle roadside cafe by the tourist path to the Falls, and there the family a little rowdily foregathered for a lunchtime fry-up, with ice lollies rummaged from the freezer cabinet, the men squinting through bluish pipe smoke; we would thereafter switch beats with an elegant old military man, always dressed in fawn corduroys and a tie, who was a tremendously well-trained killer of fish. It was not until I was fifteen that I was admitted to the magic circle of those veterans, and it was revealed just how specialised some of their techniques had become.

One of my uncle's legacies to me was his aversion to purism. His attitude to regulations was flexible, provided you proceeded with discretion. On the Long Pool one afternoon, he was test-driving a new rod built for him by the renowned Rob Wilson of Brora – this was a robust 14-footer, made of newfangled brown fibreglass, designed to deliver weighted flies. To my surprise and delight, from the pocket of his tweed coat he produced a tobacco tin containing a preserved prawn ready mounted on a trace, wrapped with copper wire and armed with two hooks, which he then carefully introduced at the head of the run, overhead casting with great aplomb (it was a strictly fly-only water, but it was evident he had done this before). 'A last resort,' he explained, though it was not successful that day.

There was another method that was illegal, underhand, unsporting, skilful and extraordinarily exciting – a tactic at which Mr Murray excelled, and which was apparently used to put an occasional fish on the table in wartime. It was called simply the Big Fly, and involved the covert practice of poaching a fish by sniggling with a blind upstream cast into its lie, and foul-hooking it in the body, instead of trying to coax it into taking the fly voluntarily – this should have felt disgraceful, but somehow it never did. Over several seasons I saw this being deployed – only ever on pools well out of sight of the road – and the illicit thrill became part of our secret, shared fascination with the river, a freemasonry binding me to silence along with the adult initiates, and which now must form part of my *confessio piscatoris*. This nearly ended in disaster one morning. I was fighting a fish hooked on a biggish copper tube and round the corner (it was in fact the Round Pool) marched the proprietor. Mr Murray hissed an oath as the laird approached and offered to net the fish for us. I was attached by the usual exceedingly heavy Maxima nylon, and could never have broken off the unfortunate fish, so I was convinced our game was up. Fearsome adrenaline sluiced through me as the salmon approached its final reckoning. 'Not too hard on him,' advised Sir John, sinking the rim of the net. Mr Murray aimed a look at me, and wrenched in smoke from his cigarette. The fish (a fresh 11-pounder) was duly hoisted aloft in the meshes, whereupon the fly was seen to be lodged fair and square in the back of its throat: against all the odds, it had actually taken the huge lure in a conventional way, and was not hooked illegally.

A more legitimate triumph came one summer's evening when I was running a Silver Stoat on a size 8 single iron down Rocky Cast, and a darkish bulge on the surface cancelled my little fly like a typing error. The fish turned tail and headed back towards the Kyle, as we stumbled in pursuit through Fir Dam, the two Claregs and finally into Paradise where, in the cinnamon light of August, Mr Murray stepped into the shallows with his gaff, and after a certain amount of bronze wallowing there emerged a big, coloured cock salmon, spectacularly kyped, squirming with the agony of that old metal crook through his speckled shoulder, a gasp wheezing from his throat, his great eye swivelled downwards as if still searching for depths into which he could escape. Though he was leanish and unhandsome, we kept him anyway, for smoking, and he weighed 22 pounds back on the Lodge scales – a notably large fish by any standards, and gaining me (at fifteen) my first ever mention in dispatches in *Trout & Salmon* magazine.

Megan Boyd was already a fly-dresser of world renown, and she lived modestly and alone overlooking the sea at Kintradwell, on the Brora road. My mother (whom Megan recognised from her films *Great Expectations* and *Kind Hearts and Coronets*) drove me to visit one morning, and left me there for a lesson – rather like dropping by on Pythagoras for a little help with your maths homework. With her severe, self-cut hair and stout brogues, Megan had a superficially masculine appearance but she had the most delicate fingers, on the deftness of which we focused for two hours each week. Her tin-roofed workshop was festooned with hides and plumes, of which she gave me a generous selection – Indian Crow, Barred Summer Duck – and amid the antique reek of naphtha, wax and varnish we sat at her kidney-shaped dressing table while she instructed me in the arts of hackling, dubbing and marrying, beginning with a Hairy Mary and eventually – after two summers – graduating to a fully dressed Jock Scott. I was a singularly inept pupil, and my tyings are still not worth two a penny, whereas hers individually fetch hundreds of pounds each to collectors at auction, and I have several framed arrays of her handiwork, including virtually the last tiny salmon patterns she ever dressed before her eyesight faded.

Over the decades she had received some bizarre requests – incorporating eyebrows, or feathers from a family parrot – and delighted in telling me about the angler who wanted a ball of

KitKat foil attaching to a hook, as he'd seen a salmon rise to take it when he flicked his elevenses wrapper into the Brora. 'There's no one more silly than a fisherman,' she would say fondly. And that's another thing I learned from her.

Years later, when Eric Steel came to interview me for his film about Megan (*Kiss the Water*, 2013) I facetiously said I thought the wing material for the original Hairy Mary Dressing had been pubic hair donated by an Inverness barmaid. Megan would have been shocked that her erstwhile student had not remembered fine well it was from a roebuck in his summer coat.

By now I was a dispirited Etonian, my schooldays an amalgam of testosterone, patchouli oil, start-up stubble and the Moody Blues. I dreamed of groupies and rock-stardom – unlikely, given my knapped visage and the fact that I only played the tuba. In my room, with its poster of Anita Ekberg among other signs of teenage unfulfilment, I had begun to compose maudlin verses. My housemaster, perusing one piece ('Dark Flower', which ended, 'Consider the lilies, dead and black/ They toil not, they don't even smell nice any more'), drawled, with a thin smile of distaste, 'I do sometimes worry about you, old thing.' 'Yes, sir.'

Ridiculed by the sporting jocks, since I had dutifully set up an elaborately stocked fly-dressing kit on my desk and was anyway regarded as a regular Caspar Milquetoast, I joined the school's angling society, and spent much time creeping around the river bank with rod and line. I began disappointing my parents in plenty of predictable ways – affecting to despise all ball games (the man who can't dance claims the band can't play), sulking in chapel, refusing to join the Corps in defiance of my exasperated father (a decorated wartime brigadier, I believe he had survived worse snubs), and spending too much time in the library. This attracted yet further opprobrium from the athletes of the House, who were perhaps less intrigued than me by the consequences of the Treaty of Unkiar-Skelessi, or the sprung rhythms of Gerard Manley Hopkins.

In July 1971, there were two further stepping stones, both on holiday with my parents in East Africa. At the Mnarani Club, Kilifi, we chartered a small tuna boat and I tried my hand at some blue-water fishing. I caught by trolling numerous bonito, yellowfin tuna and even a couple of sharks – one, a Lazy Grey weighing 118 pounds, was considered a trophy; another sheared through the woven steel trace, the frayed remnants of which I wore as a bracelet until it disappeared into the mud of the playing fields (perhaps a conundrum for some future archaeologist). This was my first experience of that crackling sensation as line streams out from a big game reel – the thunder under your thumbs that Dylan Thomas described – and though I enjoyed the bloody mayhem and onboard heroics of the trip, that branch of the sport has never quite enthralled me, despite in later years having boated a number of marlin and sailfish under similarly blue skies.

In our little safari group was a plutocratic Essex heating engineer who had in tow two wildly attractive teenaged daughters. Being awkward and tongue-tied, it was not until the penultimate evening of our eighteen-day holiday that I summoned sufficient courage to ask the elder one for a walk on the Mombasa beach (her sister seemed already to have formed an alliance with a man called Jesus, from the hotel discotheque). All went well as we strolled through the dim moonlight, until the entire sandscape seemed to shift with a general heave and a synchronised clicking, and I realised we were barefoot among a veritable morass of nocturnal crabs. Ghost crabs? Land crabs? I fear I did not wait to identify species, but, abandoning all thoughts of romance and vestiges of chivalry, I bolted for the safety of the nearest rondavel, my virtue still unbesmirched and my humiliation burningly complete.

Towards the end of his ownership of the Lodge, when ageing hips finally meant he could no longer take the hill, Uncle H. rounded off my period of sporting apprenticeship by taking me on a trip to the eastern seaboard of Canada. Having been a director of British Airways, he flew in first class as a perk, and pretended to assume that this applied to his younger travelling companion. Accordingly, with practised panache, he bypassed the check-in queue and arranged a complimentary upgrade for me (another first – I was still a student). In the lounge he cited a friend of his who, on his deathbed, had been asked if he had any regrets, replying, 'No, I can honestly say that I don't think I have denied myself anything.'

From Halifax we boarded a private Piper – Uncle happily at the controls for some of the way – and for an hour overflew tracts of forestry before landing at the airstrip for Downs Gulch camp, on the storied Restigouche river. Our host was K. C. Irving, who owned two million acres of that forestry, along with some three thousand petrol stations, having started his oil empire as a youthful gas pump attendant. He and Uncle Harold had known each other since Flying Corps days and, although he did not fish, K. C. operated this elaborate camp for hospitality purposes, and was there in person – trim, frugal, quietly spoken and teetotal (we had been forewarned the camp was officially 'dry', and there would be none of those traditional snow-ice Martini cocktails mixed with balsam needles, nor was the Admiral with us). In all, there were some 26 miles of river for his guests to fish, and it was to be just us and his family. His teenaged granddaughter Judy informed me at supper that she had never caught a salmon weighing less than 20 pounds. I slept but fitfully.

The fishing was out of narrow canoes, and my guide for the week was Wilfred Roy, a cantankerous, unshaven character in a mackinaw jacket and a seed-corn cap, who seemed dis- affected that his 'sport' was a British teenager and not some swell 'Murcan' hotshot. 'You see any bears around camp, eh,' was his first advice, 'jest throw shit in their face.' I enquired where I would find any. 'Jest put yer hand down behind yer – it'll be there.' This was not quite the badinage I had become accustomed to with Mr Murray, but following that we got along fairly well – not least because we enjoyed a bonanza.

First, sitting in the bows with his Polaroid glasses, he would drift down an allotted pool scanning the clear water for where fish might be lying up. Then, with a single-hander and a floating line, we would cover these desirable residences. The New Brunswick regulations then were that you could not return salmon, and after killing two you had to cease fishing for the day. Sometimes we had to stop by elevenses, on one occasion with salmon of 21 and 18 pounds in the boat. I also learned how to target individual fish with the dry fly – then something of a novelty, although Uncle H. had been introduced to it by Lee Wulff himself, the doyen of floating flies, and deviser of the now classic Wulff series.

There was a concentration of fish lying in a pyramid formation at the run into one deep and broad pool. I was throwing a white bucktail streamer, when the largest salmon I had seen in my life tilted gently up and turned down with my fly. For a brief while it wavered in the current like a great flag, then arrowed upstream of the canoe, shivered up out of the water once, then ran me straight around the anchor rope, popped the leader and resumed his original position. It remains one of the biggest Atlantics I have ever hooked – Wilfred put it at around 40 pounds, and now, hundreds of salmon later – a fair few of those also from the Maritimes – I still fully believe it would have weighed that much.

These days, I have my own Highland home. We live up a Perthshire glen, overlooking a loch. I have a tackle room, with plenty of rod racks, and one of those leonine cartridge bags (which tends to provoke comment at shoots). There's even a solitary gooseberry bush. Some mornings I climb the brae behind our house, with its long view of the Vale of Atholl, the high tops of the Cairngorms behind, and a haze of mist lying along the Garry river like patches of gossamer applied to staunch a wound. I owe so much of my love for these places to my old mentor, and when I visit the Falls of Shin – the cafe long gone, and a smart viewing balcony installed for the tourists – I am suffused with a bittersweet affection for that place where I first learned to read water. And in the layby I fancy I glimpse a gentleman from an earlier time, leaning against the sliding door of his Bedford van, raising his pipe stem to me in greeting. At least he knows the Admiral is in safe hands.

5

OF WALTON AND OTHER WRITERS

'The gentle art ... afflicted with a literature
containing a greater amount of undiluted bosh – to
say nothing of downright cribbing – than probably
any printed matter of equal bulk in existence.'

H. Cholmondeley-Pennell, *Fishing* (1887)

When I first looked through Walton's *The Compleat Angler* I was not much taken with it. As a schoolboy, I had been given a copy by my mother partly for the Eton connection but also because, by a further coincidence, I used to fish the River Lea, a Thames tributary that features in the book. Our Hertfordshire home was close to Ware – Walton's cast of characters was heading there, and he notes that on the wall of the George Inn once hung the picture of a Great Trout 'near an ell long' (3½ feet). I used to catch roach from the footpath, and drop-netted for native, white-clawed crayfish – before the invasive Signal variety decimated them like Agent Orange – and we lugged them home in a glass vase, to be tipped into boiling water with a little shriek of carapaces.

It was not until I spent several postgraduate years researching Restoration literature that I perhaps developed the requisite stamina to tackle the *Angler*, and began to have an inkling of why this book (reputedly the second most reprinted in the English language, after the King James Bible) occupies such a freakish position in the history of publishing, and has influenced such a plethora of writing since it first appeared, in 1653, as an anonymous, pocket-sized manual bound in brown sheepskin, priced eighteen pence. Because, although grudgingly acknowledged as a classic ('Something everyone wants to have read and nobody wants to read,' in Mark Twain's wry definition), today Walton doesn't appear to be on Main Street any more, and the mere mention of him can provoke dyspeptic reactions about anachronism and humourlessness, and one might be forgiven for thinking this once near-sacred text was no better than a trugful of old hollyhocks.

In terms of antiquity and sheer volume, piscatorial literature beats all other pastimes hands down. I am intrigued by its bloodlines. The pedigree is by hunting out of husbandry, with a goodly dollop of natural lore admixed, and it spills over into several genres – diaries, treatises, drama, verse (much of that dire minstrelsy) – a little of which genuinely draws on the springs of human expressiveness and has leached into mainstream letters. Some is written by authors who fish, others by fishermen who also write. 'Anglers read and write as no other sportsmen do,' reckoned Gene Burns, resident professor at Harvard's Fearing collection, which houses more than twelve thousand books on the subject. It seems fitting, too, that Oxford's Bodleian Library was founded on pilchards – the source of Sir Thomas Bodley's wife's fortune. I suspect that fishing and writing may share some kind of gene. Its characteristics would include a sense of brinkmanship, precision, instinct, intuition, concentration, the capacity for imaginative displacement, an appreciation of intensity, and sheer blind faith. One of the functions of language is to conjure up something that is not immediately there, and this informs much of the angler's dynamic, too.

Clearly it is by no means uniformly good. It suffers from the usual literary pitfalls – bathos, whimsy, overwriting, cliché, self-aggrandisement, lacklustre narrative, platitudinous bromides and Styrofoam figures of speech. Admittedly, a lack of freshness – that tendency towards comfortable sameyness – can afflict even the mightiest of tale-tellers; when the creator of Middle Earth was privately reading some work in progress to C. S. Lewis, the latter exclaimed, 'Oh, I say, Tolkien: not another fucking elf?' It is sometimes overwhelming to contemplate just how much jerked ink and rhinestone vocabulary has been applied to the historic subject of angling by centuries of harmless and dedicated quill-drivers and word-peckers, and some of it inevitably is tommyrot, applesauce and carpcrap (let's be honest – we are anglers, after all). We journalists are responsible for a lot of that inferior work, and I am perfectly aware that, fittingly, journalism is what wraps up tomorrow's fish and chips. At its feeblest, it merits Truman Capote's searing critique (of Jack Kerouac): 'It isn't writing at all – it's typing.'

In 1982, I was commissioned along with novelist Graham Swift to produce an anthology of fishing in literature, from Homer to the present day. *The Magic Wheel* involved two years spent reading in the old British Library, and afforded us a little perspective on this phenomenal body of work. I do not think it is of merely antiquarian interest, however: I am convinced that reading can give us a depth of field and a frisson of continuity with the past, and can quicken our appreciation of whatever it is we sally forth to endeavour at the water's edge. I am thinking here of writing that is not merely instructional, but inspirational – it is sometimes called 'devotional'. De Quincey identified the former as 'literature of knowledge', and the latter as 'literature of power' – and it is my contention that Walton's peculiar *Angler* glows with elements of both.

Working on the anthology provided the seed crystal for my own collection of books, now numbering almost a thousand, which is modest by the standards of certain other devotees. Some of these are wonderfully specialised and technical – I am fond of the learned M. Znamierowska-Prüffer's definitive *Thrusting Implements for Fishing in Poland and Neighbouring Countries* (1966), which is not an angling volume, strictly

speaking, but illustrates the way fishing is a broad church –
and others are portals into the past, such as *The North Punjab
Fishing Club Anglers' Handbook* of 1895, a strictly pukka work
with revisions by a certain Dr Cretin, of the Bengal Medical
Service. One disappointment in an auction job lot was the
Revd William Cooper's *A History of the Rod in all Countries*,
which proved to be an enthusiastic monograph upon flagel-
lation, as opposed to merely flogging the water.

But apart from these curiosa, I have a few simple touch-
stones for rating what I think of as 'pisc.lit.' The first – which
applies to any writing, I would suggest – is the Re-readability
Factor. I depended on this criterion when serving time as
a Booker Prize judge, and faced with some 26,424 pages of
new fiction to assess during the summer of 1989. If you can
positively savour something second time through, then I feel
it belongs in a special category. Next, I would prefer a book
that conveys something of the experience of time spent on the
water, in all its nuances. This is harder to achieve than one
might think, and was essential to Virginia Woolf's appraisal of
J. W. Hills (admittedly a relation of hers by marriage) when she
eulogised his prose: 'This book ... though made of words, has
a strange effect on the body. It lifts it out of the chair, stands
it on the banks of a river, and strikes it dumb.' In Grub Street,
we call that a 'sweetheart notice' – and then some. Allied to
this might be the litmus test of what a writer is like at the
sentence level, and whether the writing could be appreciated
by a 'civilian' – might it prove, indeed, to be about more than
just catching fish? Lastly, my most subjective criterion: would

I have enjoyed a day spent fishing in this author's company?
For me, Izaak Walton qualifies on every count.

Despite later attempts by the Romantics to typecast him as
an untutored rustic, Walton was already a published author
and a prosperous businessman aged sixty when his *Angler*
appeared during the Interregnum. The son of a Staffordshire
inn keeper (or 'tippler'), he had come to London as an apprentice and established himself, until 1644, as an ironmonger and
draper in the Fleet Street area – one wonders if he sold 'tack-
lings' from his shop, though it seems, remarkably, he did not
take up fishing himself until he was forty or so. Walton had a
gift for friendship, and had begun to move in the intellectual
circles of the capital, where he became acquainted with the
poets Donne and Jonson (the latter certainly an angler). He
was a devout Anglican, and later what is often described as
a 'staunch' Royalist. During his retirement years he became
known as the author of five 'Lives', or proto-biographies,
of worthy contemporaries (including Donne himself), and
though these make for dry, sermonical reading to the layman,
lacking the tartness of Pepys or Aubrey's marbled wit, they
remind us that the self-taught Walton was a man of very considerable learning.

Through Donne he met Sir Henry Wotton, former spy and
ambassador to Venice, a cultured gentleman and scientific
virtuoso who became Walton's mentor, inviting him to fish
the Thames during the 1630s. He was responsible for the definition of an ambassador as 'an honest gentleman sent to lie

abroad for the good of his country'. Wotton also gave Walton the idea of writing a little book about their pastime. There was already a vogue for didactic works about country life (horticultural herbals, horse training, housekeeping guides), and the concept of completeness was not new, either – Henry Peacham published *The Compleat Gentleman* in 1622 – but only Walton's volume has since run into over six hundred editions, been translated into Finnish, South Korean and Braille, been adapted for the stage and become a household name even to those who have no intention of ever reading it.

It was not by a long chalk the first ever book on its subject. That tradition stretches back into classical times (sometimes called 'halieutic' writing, after the *Halieutica*, a long second-century Greek poem in hexameters by Oppian) and, before the invention of printing, there were evidently centuries of manu-script compilations, a number of them monastic in origin, which recorded practical advice about fishing and included elaborate recipes for baits with arcane, semi-alchemical ingre-dients. There was also then, as today, a fertile oral source for piscatorial lore – bankside chit-chat, taproom gossip, table talk, anecdote and hearsay that encompassed all types of received wisdom admixed with nonsense and assorted Angler-Saxon attitudes.

In 1496, Wynkyn de Worde had printed *A Treatyse of Fysshynge wyth an Angle* which was appended to his *Boke of St Albans*, and has been attributed, without any firm evidence, to a supposedly aristocratic nun named Dame Juliana Berners, who may not even have existed. I like the fiction that this influential 'lytell plaunflett' was penned by a woman, as that makes her exceptional indeed in the subsequent history of halieutics. The *Treatyse* (by mediaeval times, 'angle' had become identified with the idea of the rod itself) was merrily plundered by later scribes, particularly for its playlist of recommended fly patterns, but Walton does not refer directly to the dame's work. However, he was clearly familiar with previous books by William Samuel, Leonard Mascall and Thomas Barker, from whom he creatively pilfered. It should be remembered that the concept of plagiarism per se had not yet been invented, and liberal borrowing or imitation was then considered a compliment (these days we label it research). In any case, Walton brought to his sources an alchemy that transformed them for ever.

His *Angler* does not offer much of a coherent plot, being ostensibly a meandering account of a five-day pilgrimage made to the Lea by two men who escape the hurly-burly of London to discuss sundry matters in dialogue, and are so verbose that no actual fishing gets undertaken until Day Three. It is a wide-ranging 'watery discourse', a good-natured mishmash of practical advice, reminiscence and speculation, with plenty of what the philosopher Bacon (who is cited frequently) dubbed 'broken knowledge' – that is, the cataloguing of natural wonders, however erroneous or incredible. It breathes the bankside immediacy of a day in the outdoors: sheltering from a shower beneath a honeysuckle hedge, secreting a bottle under a sycamore's roots, locating a specific chub with 'some bruise upon his tail'. His style, though turbid in stretches, is for the most part clear-running and easy, as with these observations on barbel: 'He is able to live in the strongest swifts of the Water, and in Summer they love the shallowest and sharpest streams; and love to lurk under weeds, and to feed on gravel against a rising ground, and will root and dig in the sands with his nose like a hog, and there nests himself.' It is a labour of love, truly the work of an *amateur* – engaging, if not always elegant, garrulous, opinionated, occasionally gullible, quirky, aphoristic, incorrigibly digressive, often deliberately archaic, cranky and cumbersome in places but suffused with a magpie curiosity and emanating a distinctive personality (which I suspect accounts for its longevity): 'the whole discourse is a kind of picture of my owne disposition', writes Walton, though I believe he is being partly disingenuous.

The bluff, peaceable, naïf yeoman persona of Walton (as channelled through his alter ego 'Piscator') is in fact an underrated feat of literary impersonation. Even by the standards of his own wistful times – a world turned upside down, with civil wars (in which 3 per cent of the population perished), regicide, disastrous harvests and all the turbulence of blood-bitterness and betrayal that characterised the Great Rebellion – Walton's inclinations are carefully and manifestly archaic. His book harks back to a golden age in Elizabethan times, and is shaded with a vision of Old Albion improbably bucolic and Arcadian, a richly plaited version of pastoral bedecked with swains and milkmaids, ballads around the wall and lavender hung in tavern windows. It is a place of sanctuary, and the spirit of the *Angler* is one of escapism during times of acute turbulence. In its ruffled contemplation of serenity, it becomes a spell against turmoil. His balmy, sunlit Maytime scenes were designed to console those who had undergone a recent succession of especially harsh winters, or whom politics had forced to retreat into the plundered countryside – though Walton knew as well as anyone that the villages of England were not all culverkeys and cool ale, nor the meadows perennially woven with bird calls and stippled in soft light, that the lanes were claggy with nightsoil, and infant mortality was rife (seven of his own children predeceased him). Yet his hymn to the virtues of a simple, good life – though it may smack of Hobbiton to the jaundiced modern reader – was sung against the forces of disorder, endorsing the belief, as his friend Bishop Aylmer would insist, that 'God is English'.

In fact, far from being 'just' a book about the recreation with which it has become synonymous, the *Compleat Angler* is a mildly subversive work. The action, such as it is, begins on May Day, a festival long associated with Lords of Misrule and fertility trysts and which caused the Puritan regime to outlaw the maypole in 1644. Walton quietly cocks a snook at the authorities, at a time when even nostalgia could be regarded as ideologically suspect. His allusions to the 'Brotherhood of the Angle' are also covert references to the embattled Anglican faithful (similar puns feature in the correspondence of his friend Gilbert Sheldon, a noted barbel fisher who later became Archbishop of Canterbury); they were continually harassed at worship by Cromwell's troops, and many of them had in the interests of safety swapped their sophisticated urban lives for exile in the country, a milieu Walton seeks to reassure them is not merely a place of empty banishment but a source of solace and restoration. His stressing of honesty and good fellowship, perceived as tedious by some today, was not just pious claptrap but an attempt to rebalance the humours in the ailing body politic, the antidote to a state of affairs where neighbours were still literally at one another's throats.

In 1676, when Walton was in his eighty-third year, there appeared a fifth (and principal) edition of his by then celebrated book, with a second part (mostly about fly-fishing) contributed by Charles Cotton, some thirty-seven years his junior. The title page distinguishes between 'Mr. Izaak Walton' and 'Charles Cotton Esq.', and on the face of it, this was an unusual pairing: the younger man, whose father had been friendly with Izaak, was a raffish and headstrong roisterer, a ribald, bibulous gentleman beset by creditors, a modish but scurrilous poet, and an expert at billiards (he had published *The Compleat Gamester* in 1674). You might not have wanted him as a son-in-law. Together they fished that charming limestone stream the Dove, near Cotton's ancestral Staffordshire home, and there he had built for 'Father Walton' a fishing temple you can still visit, their carved initials intertwined and the inscription 'Piscatoribus sacrum' commemorating their companionship. Unlike the fastidious Walton, his protégé claimed to have dashed off his contribution with gusto in ten days flat (not a bad rate for 25,000 words, take it from me) and some feel its brio appeals more to the modern angler, not least his sage advice to address trout 'fine and far off'. Walton seems primarily to have been a bait-fisher; Cotton once hung up one of his mentor's flies, 'in his parlour window to laugh at'. He only survived his elderly friend by three years – there is a memorial plaque to Cotton by the bell-rope in St James's Church, Piccadilly, and each day en route to the library while working on this book I would invariably touch it for luck.

Not all his contemporaries were enchanted by Father Walton. Richard Franck was a choleric former Cromwellian cavalry officer in the New Model Army, who encountered Izaak just once, and it seems to have made him angrier than a junkyard mastiff with an ulcerated scrotum. Politics doubtless came between them, and the Roundhead was probably miffed by the publishing success of his perceived rival (stranger things have happened). In his own logorrhœic book *Northern Memoirs* – which did not see the light of day until 1685 – he castigates Walton as crassly ignorant, 'a scribbling putationer ... deficient in Practicks, and indigent in the lineal and plain Tracts of experience'. God's hooks! Setting aside his snarling minutiae and the alarming steampunk timbre of his prose – 'Beautiful Bohanan, besieged with bogs, and barracadoed with birch-trees' – his travel book offers the first ever, genuinely adventurous, printed account of fishing in Scotland (where he went 'to rummage and rifle her rivers and rivulets'), plus the earliest description in English of the now-extinct burbot. Descending at times into *obscuranto*, freighted with overwrought anecdotes and garlanded by turgid philosophical *pronunciamentos*, a little of Franck's bravura prose goes a long way, and today, despite a valiant edition by Walter

Scott, his considerable expertise with rod and line is all but forgotten.

Izaak Walton died in Winchester in his ninetieth year, during a great frost. In his will he left funds to provide wintertime coal for the poor of Stafford, and a tantalising item: 'Fishing-Tackle and other Lumber', valued at ten pounds (just imagine its market value now). His reputation may have fluctuated down the years, but the *Angler* has not been out of print since 1759, when Dr Johnson promoted an edition. Old Izaak was subsequently hijacked by the Romantics, as the zeitgeist of a bucolic England that was already disappearing from view beneath the Industrial Revolution: Wordsworth, a keen fisher but never my favourite lyricist, waxed typically winsome, claiming Walton's *oeuvres* were written with a quill 'Dropp'd from an angel's wing'. Such hagiography provokes an inevitable backlash. Once the unofficial patron saint of the pastime, Walton's reputation now stands low in today's Car Boot Sale Britain, with its fetish for bullet points and 'elevator pitches' and an engrained antipathy towards the quaint and bosky. He may indeed be a relatively minor writer in the greater scale of literary accomplishment, but most of us scribblers would settle for a smidgen – a minnow's gleam – of his enduring fame.

It is not quite as easy as one might think to compose a 'watery discourse' that has truly durable qualities. If my pantheon of chosen authors is overarched by Walton, its floor would be inlaid with bright shards of those who produced books I'd have given my eye teeth to have written.

As well as being a popular novelist, Arthur Ransome was the doyen of fishing columnists, contributing over 150 articles to the *Manchester Guardian* in the 1920s, to which he was also the long-standing foreign correspondent. As a regular columnist myself since 1987, I am aware what a challenge this can prove, and admire his resistance to dogma, and his sonofabitch honesty of approach. He had no interest in appearing charming, or trying to cast that little touch of moonshine onto the page. As with so many skills, the trick is to render your hard work invisible. It remains a loss to the *bibliotheca piscatoria* that his Waltonian novel *Piscator and his Phillida* never materialised, and that his wife Evgenia discouraged him from ever completing his other fictional project, *The River Comes First*. The Elver fly pattern of which he was so proud was inspired by observing wartime GIs chewing gum in the lobby of the Park Lane Hotel.

Indeed, considering its penchant for make-believe, fabulation and the picaresque, there has been remarkably little piscatorial fiction. In Britain, the landmark piece is arguably Henry Williamson's *Salar the Salmon* (1935), which, if not precisely about angling, is focused by the world underwater, and handles the trick of anthropomorphism as adeptly as he had already done with Tarka. The reputation of this prolific novelist suffered from his political affiliations (he was briefly interned in 1943 for his fascist sympathies), although his nature writing about the West Country remains peerless. Fishing crops up in the fiction of John Buchan, George Orwell and others, but it is really to North America that one turns for imaginative writing that is purposively predicated on the sport.

Although not classically about angling – there are barbed 'irons', but the quarry is darted rather than duped – Herman Melville's astonishing novel *Moby-Dick* (1851) is arguably one of the greatest accounts of the 'quest by water' and is squarely in the spirit of fish hunting. Its narrator Ishmael refers insistently to the sperm whale ('parmacetti') as a fish ('I take the good old-fashioned ground that the whale is a fish') and many a subsequent angling author has had to navigate in its wake. Not least Ernest Hemingway, who may be decried for his machismo and intolerance – there are some who believe his posturings are mutton dressed as ram – but he was unbeatably evocative of the blue-water sport (fishing experienced as big game hunting) and, on a smaller scale, his Nick Adams stories remain for me the deftest of the genre. One reason American outdoors writing enjoys an expansive resonance sometimes lacking in Britain is that it has the advantage of wilderness (for example, Wyoming is slightly larger in extent than the United Kingdom but is home to some 600,000 people, as opposed to sixty-six million). There is also a continuing sense of brinkmanship and exploration, including a familiar tradition of canoe trips with Pop, and more than a breath of Thoreauvian transcendentalism out there in the backwoods. This surely informs that modern masterpiece *A River Runs Through It*, Norman Maclean's translucent depiction of his family interacting through the medium of a shared passion, one of the few examples of piscatorial fiction that makes plausible the much vaunted idea of the aquatic epiphany. Maclean died in 1990, just before filming began on what is known simply as 'The Movie' – I cherish the tale of him informing a Hollywood shyster, 'When we had bastards like you out West, we shot them for coyote bait.'

Another subtle and versatile Western novelist is the great Thomas McGuane, whose early bonefish bum Key West fiction *92 in the Shade* (1973) he later directed as a film, starring Peter Fonda. In fact, McGuane's non-fiction collection *The Longest Silence* (1999), with its tremendous sprezzatura and deep-brewed responsiveness to all things fishy, is probably my number-one pin-up among contemporary halieutics: whereas most exceptional writing on the subject urges one on to attempt better things oneself, that volume – laconic, lyrical and wise – makes me feel like chucking it all in (when he wrote an introduction to the book of one of my chums, I became as bitter and twisted as an old billfish leader).

The panoply of American authors to admire and emulate must also include Zane Grey, a Pennsylvanian dentist who turned to fiction, selling in his lifetime fifteen million copies of his (mostly Western) novels on the proceeds of which he fished around the world, though his competitive attitude to the sport made Papa look like a mere trainspotter. Grey patronised the younger Van Campen Heilner, sybaritic heir to a mining fortune who was also a field representative for the American Museum of Natural History, and in 1937 published *Salt Water Fishing*, an instant best-seller that is sprightly and quixotic, pioneering a spectrum of maritime methods and yet witty enough to be read by the uninitiated. My whistle-stop survey of that continent would also include Roderick Haig-Brown (who, though born an Englishman, came to encapsulate the Canadian wilderness), the encyclopaedic, globetrotting A. J. McClane, the stainless steel sharpness of satirist Ed Zern, urbane and elegant Nick Lyons, the erudite Ted Leeson and the miraculously re-readable and lucid Bernard 'Lefty' Kreh, modest despite the profundity of his knowledge. Finally, the prolific and always surprising John Gierach seems to have no European equivalent – his prose is a sassy blend of hip, poetic and sardonic, a mellow combination of gravity and levity which several pale imitators have failed to achieve.

Back in Europe, I repeatedly enjoy the company of three authors who now smack somewhat of yesteryear, but with whom I feel sure a day with rod and reel would have been congenial and enlightening. Patrick Chalmers was an enterprising poetaster whose greatest achievement was *At the Tail of the Weir* (1932), an elegiac biography of his beloved Thames in the spirit of Walton, the phrases soft as snowfall and just as bright. His description of barbel singing in the moonlight is typical of the whimsy and gentleness of this historic volume. H. T. Sheringham was an unusually good-natured, droll enthusiast who (like Chalmers) was an expert 'coarse' fisherman as well as having been brought up to pursue those 'game' species of the gentry – he was thus a member of an endangered species, the all-rounder. His journalism was widely published, and of his books I think I prefer *An Open Creel* (1910) which is nicely meditative and ironical. And last, that quietly percipient Irish Supreme Court judge, T. C. Kingsmill Moore, whose memoir *A Man May Fish* (1960) is a gentle and civilised celebration of fly-fishing by a modest and trenchant practitioner – a day on Lough Corrib with his wily old boatman Jamesie Donnellan would have suited me just fine.

There is another style of piscatory prose less serene, of course, and it finds expression in the books of Anglophile American Negley Farson – chiefly his paradigmatic *Going Fishing* (1942), an unsentimental, economic account of his travels which chronicles an edgy lifestyle with an exceptional eye for vignette. Farson, who was seriously wounded in the Great War, was a heroic smoker, and also a drinker of some diligence (his wife Eve suggested the book be entitled 'With Rod and Gun Down the Alimentary Canal') and his hard-bitten approach to vicissitude is married to a beady apprehension of the natural world. He spent time both in sanatoriums and the veldt, but in 1960 this swashbuckler died in a wheelchair with his dog at his feet.

Mentoring has played a key role in the evolution of our sport (though it's becoming rarer), and while I can't pretend I became a C. Cotton Esquire to either of them, those twin colossi of modern angling, Hugh Falkus and Fred Buller, both at different times took me under their wings. Falkus was a man in the Farson mould, and was a great divider of opinion. He was an enigmatic combination of shaman and showman (with a touch of sham), roguish, insightful, narcissistic, charismatic (with a burning blue gaze and vatic white locks), uncompromising, dogmatic, a perfectionist, a bearer of grudges, brilliant on the page. In a drinkative disagreement late one night, he came at me swinging a forearm as big as a bacon flitch, yet his books (of which the childhood memoir *The Stolen Years*, 1965, is probably the finest, though he is best known for *Sea-Trout Fishing*, 1962) are clear and bold and enduring. He once rang me after a skinful of High Commissioner Scotch and growled, 'My dear sir, I want you to know that your latest book is a complete and utter load of crap.' His refrigerator may not have been stocked with the milk of human kindness, but Old Falk was a remarkably powerful communicator.

Fred collaborated with him on the excellent *Freshwater Fishing* (1975) and was the most impressive angler I have known, steeped in the history of the pastime and with an unassuming, though firm manner. He would never have stomped off and left me in the deep dark on a riverbank, as Hugh did when my steeple casting wasn't up to scratch on his stretch of the Cumbrian Esk. I admired Fred for his methodical mind and his engaging, scholarly style – he seemed not to suffer from 'the delusions of omniscience' which afflict many fishing writers, in Ed Zern's astute phrase. I was with him on a lake in July 2015, when he enjoyed his last ever day's fishing.

Set against such exemplars of eloquence and conciseness, much of our slack modern writing, with its screaming reels, bars of silver and other formalin-injected clichés that have literally become our catchphrases, seems to emit but a feeble voltage. Yet even the deftest of authors can be floored by the particular demands of writing about a specialised pursuit:

reporting on a horse race as his first assignment for *Sports Illustrated*, after an hour all the novelist Kurt Vonnegut Jr could come up with was the sentence, 'The horse jumped over the fucking fence.' Ah, the Romance of Words.

6

BORDERLAND

'"How deep should I go?" said the enterprising
man. One said to the fifth button of your waistcoat,
and the other to your shirt-collar.'

William Scrope, *Days and Nights of Salmon Fishing*
in the Tweed (1843)

It was the year I had my wisdom teeth out that I caught my
first Tweed salmon. Being a teenager with a morbid imagination, I had asked the surgeon if I could keep them as
souvenirs, and as I stretched out for a post-operative drink
of water I brought up to my lips, instead, the glass container
with my bloody teeth in it, tendrils of skin still attached. As
a convalescent treat, I had saved up for my first Hardy spinning rod, and an Ambassadeur reel, and with this I landed an

8-pound springer on a brown and gold Devon minnow from the Shot pool on the Upper Floors beat, after which my noble host served up plovers' eggs and oriental salt for tea.

Since then, it has become the river I know best, having fished fifteen of its beats – I admire the Dee and Spey, but Tweed (the definite article is customarily not used) is the stream with which I have established some sort of rapport. Memory may well furnish the fourth dimension to landscape, and I have been haunting its banks now for almost fifty years. It is also arguably – after the Thames – Britain's most literary river.

My uncle Harold had lived in the Borders at Hendersyde Park during the 1950s, and each October he would rent a week's fishing at Upper Mertoun, further upstream near St Boswells. When I graduated in 1977, he asked me to join him there for two days, and so began an association with that beat which was probably the most spoiling in all my life. He shared the fishing with a plutocratic butcher from the Home Counties, who would be driven from the hotel each morning in his Rolls, enjoying the first of his four daily Havanas. His chauffeur-cum-factotum, Harry, sported a thin moustache and was as nattily turned out as a pox doctor's clerk; he was famous for his compendious collection of printed pornography, but he also loved to fish and was allowed an hour on the bank when the gents were lolling in their lunch hut. The boatman then (there are no 'gillies' on Tweed) was a leathern-faced veteran named Jim Fletcher, who still wore canvas waders, and sang Borders ballads in the bar, and harked back to his early days with, 'Fifteen feet o' greenhairt, and the reel line runnin' free.'

We always stayed in Kelso, at the Ednam House Hotel – a large, family-run place of faded splendour with fringed lampshades and ornate plasterwork (the style was perhaps Late Antimacassar period), where the food was consistently erratic, and around six in the evening a maid would knock at your door and offer to turn your candlewick down for you. The night porter was melancholy and had a gammy leg. With the ambience of an informal club, it hosted virtually the same guests at the same tables every year – a curious core sample of the well-heeled at leisure, including a contingent of retired military types with tightly permed spouses in quilted gilets gripping the *Daily Telegraph*, plus a skein of retired wool manufacturers and civic worthies from Yorkshire who turned each day into a party, starting at breakfast. One of them had also been a pilot in the RFC and gave my uncle a tobacco pouch to match the one he had made himself in the Corps colours – these two survivors used to wave them at one another before lighting up in the evening. As we descended the stairs to dine, Mr White would already be seated to the left, at his own separate table, hands neatly folded in anticipation of the nightly ice cream with strawberry syrup.

At that time, the prodigious runs of back-end salmon were reaching their zenith and when you returned from your beat the hotel hall would be paved with silver – great, slab-sided fish, surely a race apart from their streamlined summer cousins. Sometimes it looked more like a Chicago-style drive-by shooting than the results of one autumn day on a river. Back then, we knocked almost everything on the head (the coloured specimens would be sent straight to the smokery from the boatman's freezer), to the astonishment of visiting American anglers, who had already developed the practice of 'catch-and-release', and must have thought we shared the motto of the Spanish Foreign Legion: *'Viva la Muerte'*.

The cyclical appearance of those same 'well-kent' faces each October was peculiarly reassuring, like the bountiful numbers of the fish themselves. In the bar, it was as if the circus had come to town: a perilous fug of smoke and Scotch, a mêlée of boatmen, cricket stars, Lords of Misrule and piscatorial pundits jabbing forefingers to emphasise theories, and a number of shonky characters who were often all hat and no cattle. You fell among thieves, into the early hours. There were occasional scuffles. Next day, you awoke with a Minotaur head and your mouth would taste like the inside of a Pathan tribesman's food wallet, necessitating (around elevenses) what Cyril Connolly called 'a tuft of the dog'. I may have courted oblivion on a nightly basis, but I loved almost every carnival minute of it, year after year. Truly, it seemed like a golden age.

For 17 miles of its long course, Tweed forms the actual border between Scotland and England, and this region once comprised the Debatable Land (Berwick alone has changed hands fourteen times, and it may not be over yet). Though subsequently romanticised, this disputed territory was for centuries a bloody threshold for reivers, moss-troopers, assorted villains and local heroes – the chief of the skirmish-ing Ruthvens was known as the 'Earl of Hell'. No other river in these kingdoms carries such a penumbra of past violence and mystery, which – particularly in the middle reaches – imbues with a frisson your modern attempts to harvest it with a puny, sweeping fly-line.

Watling Street ends nearby, but the remote Cheviots were an unpopular posting for Roman legionaries – the Libyan-born Emperor Severus lost fifty thousand men here in just three years. The Picts were a wiry, river-revering race: it is said the two last survivors (a father and son) were captured by the Scots and refused to divulge their secret recipe for heather ale, which died with them. Then came Arthur, waging war against the pagan Saxons and now lying asleep beneath the Eildon hills (opposite Mertoun), with his retinue of white horses. There were two Merlins hereabouts – one from Camelot, and another known as Merlin Caledonius, a sixth-century warrior bard who was clubbed to death by fearful shepherds and is buried under a green mound (the Wizard Thorn) by Powsail Burn. His prophecy was fulfilled when it overflowed and mingled with the water of Tweed at the union of the crowns in 1603 – maybe it will flow in reverse, if Scotland regains its independence. Stranger things have happened in these Borderlands.

Thomas the Rhymer – historically Sir Thomas Learmouth, laird of Earlston in the thirteenth century – became a seer following seven years spent in thrall to the Queen of Elfdom (also beneath the Eildons). The hills themselves were reckoned to have been split in three by a demon at the behest of his contemporary Michael Scot, the magus of Ettrick, a scholar of chemistry and Arabic, author of *The Secret of Secrets*, who acquired uncanny wisdom by cooking a white snake and sucking his scorched finger. He killed his wife, a rival enchanter, by putting heated crows' eggs in her armpits, which apparently no practitioner of the dark arts can survive, and met his own end when felled by falling masonry during Communion. Physically buried in Melrose Abbey, his spirit languishes in Dante's purgatory. Small wonder my hands shake and the bloodstream fizzles as one casts beneath such shadows.

From its tundra-like uplands (the 'grey hills' where John Buchan, later Lord Tweedsmuir, was raised) through the middle reaches, or Garden of Tweed, and down to the broad, fertile pastures and haughs, the river runs for almost 100 variegated miles; from high up in Tweeddale (near Arthur's Seat, and the Devil's Beef Tub where rustled cattle were secreted) it was once said you could see into nineteen counties. For years its power has been used for mills and channelled with caulds, a few of which remain and still prove ideal holding places for fish. The textile industry, which burgeoned when Sir Walter Scott popularised the vogue for Border check patterns in the 1830s, still produces tweed, though the name of the cloth owes nothing to the river – it is a ghost word that arose from a clerk's misreading of an invoice for 'Scottish tweels', or twills.

Some folk feel that autumn marks the tarnishing of the year, but on the river here it has always seemed a time of delicious anticipation – those October mornings when the frost-shrunken current rattles beneath sandstone scaurs the colour of rhubarb fool, there is spiderling gossamer in the air, the sepulchral smell of damp bankside moss, and the stands of birch, ash and elm are still swagged with foliage, the slow fires of seasonal colour scorching through them before cold winds spit them into the water, when for a while you hook one on every other cast, as your sunk line bites into the brisk stream. But Middle Tweed was not always so gladly wooded. When Washington Irving visited in 1817, he described the banks as bare. It was Scott (the Wizard of the North) who was partially responsible for planting out this area, offering a penny to any child who brought him an acorn.

The Great Romancer tended not to settle for half measures. When he conceived his Gothic monument of Abbotsford, it was just a boggy place named Clarty Hole. Faced with financial ruin after his publishing venture collapsed, he wrote his way out of adversity, timing his daily output against an hourglass, and produced the hugely popular Waverley series – novels now mostly bought by interior decorators needing to furnish by the yard the shelves of non-reading clients, but once so influential there were twenty-two towns named Waverley in North America (his *oeuvre* also inspired Heart of Midlothian football club). It is said he rediscovered his notes for the project in 1813, while searching for some fishing tackle to lend his house guests, and, though not much of an angler himself, he certainly understood the culture of the sport. In *St Ronan's Well* (1823) he has his snobbish members of the

Claret Club concede that one newcomer might be a gentleman because he could cast 12 yards of line single-handed and 'the fly fell like a thistledown on the water'. It may be that his lameness from childhood polio made wading difficult for him, though he enjoyed the spectacle of 'burning the water' (spearing salmon with leisters by the light of flambeaux, which features in *Guy Mannering*, 1815), and occasionally went after 'caller' trout – coloured, as opposed to 'siller' (silver) – in St Mary's Loch with his poetical collaborator James Hogg, the shepherd, and assorted Edinburgh literati whose breakfast comprised 'parritch, wi' maybe a thimblefull of brandy'.

A regular guest at Abbotsford was the chemical genius Sir Humphry Davy – a keen fisher from youth (when he was bitten by a Cornish conger), who designed his own green fustian angling outfit. Davy knew the area because he had married a wealthy but tiresome widow from Kelso. He wrote a volume of tortuous dialogue, *Salmonia* (1828), about his passion, and died of apoplexy on a trout-fishing tour of Switzerland. John Younger, the radical, autodidact cobbler poet of St Boswells, whose fishing brogues cost five guineas a pair to buy, reckoned Davy's book was 'something very like nonsense'. He resented the way Scott and his circle were attracting southern sportsmen to Tweed, raising the rents and ousting the locals from their accustomed sport (a sentiment not unknown in other parts of Scotland). Scott died in 1832 on a warm autumnal day, the window open so he could hear the river – 'the sound of all others most delicious to his ear'. On the way to his burial at Dryburgh, the horses drawing his hearse stopped instinctively at his favourite vantage point, now Scott's View. He never saw the legacy of his lovely treescapes.

The tenant of Mertoun, William Scrope, was another of Younger's bêtes noires – a stylish, multi-talented Victorian classicist and pioneering deer stalker with a penchant for snuff, who gave not a *foutre* for convention. He had introduced his friend Sir Edwin Landseer to angling, who then provided some of the illustrations for Scrope's quirky, outstanding book (from which the epigraph to this chapter is taken). It was written at Mertoun where he had the lease for seven years. For a guinea a week he also hired the boatman John Haliburton to help him rack up the tally of fish taken, which in those days legally included kelts (Younger claimed Haliburton hooked many of Scrope's salmon). The other celebrated professional of the period was Rob o' the Trows (Robert Kerss of Makerstoun, an expert much in demand). One day he was boating a 'nobleman' who killed three fish, celebrating the demise of each with a dram poured for himself alone. Exasperated, Rob chained up the boat and departed, saying, 'Aw's just thinking that if ye drink by yersel, ye may e'en fash by yersel.'

Scrope was an intrepid sportsman, wading eight-hour sessions with his 20-foot fly rod from Higginbotham in the Strand, periodically emerging to see if his legs had yet turned black (up at Melrose, St Cuthbert used to wade naked all night in winter to mortify his flesh – one can only wonder what colour his legs must have been). One favourite fly pattern, the Lady of Mertoun, required its body to be made from spun water rat's fur, and another (Meg with the Muckle

Mouth) was named after the unprepossessing daughter of Sir Gideon Murray, who was eventually married to an ancestor of Scott's when forced to choose between that fate and the gallows tree. When fishing this beat, then, I sometimes stood on the selfsame boulders as one of the sport's greatest enthusiasts, perhaps even hanging my lure over a similar lie, or lodge, where white water jumps up before the tail of a glide, the entire experience subtly cross-hatched by the quick and the dead.

Eventually, my father joined the Kelso week, but during my teenage years we had tried out some summer holiday venues where we could spend quiet time together, turning the thread that would bind us together more closely as adults than my childhood had ever allowed. The drive north itself became a ritual – conversation is sometimes easier when you don't make eye contact. His cheerful insouciance about anything technical concerning the sport ensured that we enjoyed our trips even if they proved unproductive – I doubt my father (unlike his son) ever thought about fishing when he was off the water: what he relished was the social side of being away from home.

For two August weeks in the late 1970s we stayed at Loch Maree in Wester Ross, at its historic hotel girt with rhododendrons and blurred with midges, where the proprietor was a stern and robustly built matron named Miss Moodie. Regular guests were treated deferentially, 'very much as the Hotel Meurice used at one time to regard the Emperor of Trebizond and the Vidam of Pschmskp', as that witty essayist William

Caine put it, of another bygone inn. Although this was not in their brochure, the hotel was infamous for having been the site (in 1922) of Britain's worst outbreak of botulism, when potted duck paste in the packed lunches killed eight; those collecting their picnic baskets late were given ham, and survived.

The spectacular loch holds twenty-seven islands, many of them wooded with pine – one (Gubhaim) contains its own lochan with, in turn, an islet upon it, and Eilean Maree, a place of pilgrimage since Viking times, has an emaciated good-luck tree embedded with votive copper pennies. Loch Maree was famous for its run of big sea trout – rods accounted for a couple of thousand annually, until fish-farming operations began in 1987 and put an end to it. The principal method was dapping: this involves no actual casting, but a light silk floss line billows in the breeze from the top of your rod and you work a bushy fly along the surface. Some find this tedious, although to me the blow-line is intriguing, with a need for concentration and holding your nerve, particularly when provoking jacuzzi-swirl rises. Kenny McKenzie, our gillie, was rumoured to speak Chinese, and had a wooden leg. Jumping trains down in Patagonia as a young man, his coat had snagged a door and he was dragged beneath the wheels, returning to spend the rest of his days as the Captain Ahab of Gairloch. He arrived at the jetty each morning on a specially modified motorcycle. There was a colleague, Willie, known as Willie Nelson, since he had lost an eye in the Great War.

We never did hit the right conditions for success, being frequently becalmed and rarely seeing the water curling and gleaming under that westerly breeze needed for long drifts across beats such as Steamer Channel or Fool's Rock. We only managed finnock, mostly returned. It was the practice of Donald the porter to display guests' daily catches on white ceramic trays in the hallway, and on our last evening we laid out two empty miniature Bell's whisky bottles with a label stating for posterity, 'Profumo & Profumo. Ash Island.' The hotel closed its doors in 2007.

Another of my Tweedside heroes was Thomas Tod Stoddart, eccentric scion of an Edinburgh admiral, who in 1837 foreswore a legal career to settle in Kelso (with his wife, Bessadh, a crofter's daughter who only spoke Gaelic) and devote his life to fishing the Borderlands. He was described by one friend on the banks of the Teviot 'literally clad with salmon and sea trout', and sometimes the weight of fish, coupled with 9 pounds of leather waders and boots, on top of chronic lumbago and sciatica, constrained him to stop early. As a boy, he had learned about a deadly bait called 'patt', from Monsieur Senebrier, a former Napoleonic prisoner of war. Patt, then a legal paste, was concocted from salmon roe and was so attractive that the trout would be nibbling his fingertips as he rinsed them in the shallows at the end of the day. (I have met at least one boatman who still knew the recipe, but have never dared try it. Maybe that was The Secret of Secrets.)

Our years at Mertoun were never quite in that league, but we had a number of red-letter days. On one, I landed 94 pounds of salmon to my own rod. The water was in perfect ply, dropping and clearing after a weekend rise, the fish were settled and the boatman held me with strength and dexterity over a series of desirable residences for hour after hour, my WetCel 2 line and Gordon's Fancy tube, cast square and mended, working down the tail of the Cauld Pool all morning, then from the left bank of the Bridge in the afternoon. Such an alignment of ideal conditions and great good fortune rarely lasted long, and the honours were seldom even. My father-in-law, Alasdair Fraser, a Highland Scot and an exceptional sportsman, once had four while I skunked (it was almost separate tables at lunch; but I was pleased when in the hotel that evening a chap celebrating the Kelso sheep sales lurchingly asked him if he'd fetched good prices for his tups and hoggs, an enquiry he had never received in all his years as a Harley Street gynaecologist).

Inevitably, there were drought years when the shrivelled stream seemed to glare at you from its slimy bed of stones, and the whole of Tweedside could have done with a thorough detox. 'I've seen more life in the Dead Sea,' would be boatman John Taylor's lugubrious remark. In 1994, it was at its lowest for eighty-two autumns, but I managed one tired kipper from the lip of the cauld; I suppose it is really the minor triumphs, as opposed to the bonanza days, that define the spirit of salmon fishing, though at such times it effectively becomes the Art of the Impossible. As a party we only ever blanked in 1989. Some weeks were washed out by dirty water rising, then developing into an Augean spate ('ower drummly', in Scrope's vernacular) that shuddered through the beat for days on end, chocolatey from Ettrick water or reddish from the Leader, rendering the residents flood-sick, bringing down straw bales and swollen farm animals and an entire summer's worth of bankside detritus. You have to pay your dues. The Victorian Lord Hume used to reckon, 'In all my long experience of salmon-fishing, I do not think I have averaged more than two really good days a year' – by which he may have meant occasions on which he was positively clad in salmon.

Sometimes the fish would just lick your fly, with a 'pook', and you could not buy a proper take. John kept my spirits up when that was happening on the fabled Back Braes by telling me of one gentleman who was pulled twenty-seven times on that pool in one morning, without a single offer being converted, but during the very next session his wife landed seven there on the trot. *Es la pesca*. Or, as my own wife Helen remarked verbatim, when she grassed her first salmon here, 'Well, if you're so good, how come you didn't catch one yourself?' I am still groping like a guddler for the answer to that.

In 1990, during high water, I landed a fresh hen fish weighing 28 pounds, my largest ever, the photograph of which hangs in the Rogues' Gallery of Ednam House. But as so often, rummaging in the lumber room of memory, it is the losses one recalls (my theory is that the number of salmon an angler loses is in direct proportion to his misdemeanours in a previous life, in which case I must have been alarmingly debauched). In 1979, the boatman Lackie and I had killed three nice salmon by four o'clock, duly wetting the head of each from the bottle of Pig's Nose Scotch in my tackle bag. It was breezy, and by now my casts were going all over the place, like a madwoman's custard, but at the neck of the Braes, just at that sweet spot where two conflicting currents clash, I was taken by a fish – the loop of line was drawn from between my forefinger and the cork grip as I fed it into my customary slip-strike. I came tight to him, the rod duly described its battle curve, and a dorsal and caudal fin surfaced, an impressive distance apart from one another. Then the salmon lugged away and kept deep. After such a triumphant day, I was fool-ishly cavalier with this latest encounter and began horsing it across the gravel channel (we were using heavy nylon, which in the old days would have been called 'elephant gut'), but when it approached the boat, head still down, and we saw the size of its fuselage (I refuse to call salmon 'a bar of silver', as they simply are not ingot-shaped), Lackie said, 'Christ, Mister David', followed by a further curse that almost burned the beard off his face. I had been too hard on it. The hookhold gave way, my rod straightened abruptly, and a fish that must have been well over 30 pounds – a great fish that I had never deserved – dissolved back into the deep darkness from which it had briefly been summoned.

I often fished the bank rod, as there were only ever two boats in action and I was usually the youngest in our party. The Mill Stream which churns from below the cauld was sometimes thronging with fish, and alone one noontime I am vigorously engaged in a protracted dance of death up and down its rubbly banks with a remarkably sizeable fresh fish (a *monstrum horrendum ingens*, in Scrope's words), when Mr White cruises up in his black Sierra, bringing provisions to the nearby luncheon hut. It is a tough place to land a fish, so I ask if he would help me by fetching the heavy net propped against the wooden fence. In a natty anorak and those Chelsea boots, he totters down to the water's edge and strikes a pose like some amateur actor impersonating a harpoonist in the prow of a whaler. I bid him do nothing until we are ready, and he peers uncertainly into the margins, never before having been involved with such a drama. The fish turns on its side, I begin to drop her down the eddy, and then, with the noblest of intentions, Mr White executes an extravagant pass with his net, which comes back up with only the Waddington hooked in its meshes, torn from the jaws of the flailing salmon. 'Ooh, it's gone.' He looks stricken. 'Yes, yes, I know.' I light a Senior Service, and wait for my heartsickness to leach away, like the smoke whorling from my mouth.

The Royal Burgh of Peebles has had as its motto since 1473, '*Contra nando incrementum*', emblazoned with three salmon – 'growth by swimming against' – and for centuries the nomadic lifestyle of this mysterious migrant has snagged the popular imagination, inspiring much metaphor and myth. It's not hard to see why, as *Salmo salar* is one of the few fish to have the charismatic appeal of certain rare birds, or some of the terrestrial megafauna.

After thousands of miles at sea, he comes forging up from the dark tide towards his natal headwaters, navigating somehow by starlight and gravity, using his miraculous sense of smell to home in on the particular *terroir* that spawned him – the clan chief of the Salmonidae, with the vigour of the deeps still upon him, anadromous, philopatric, euryhaline, a lustrous creature from our ancient past, dwelling between two worlds, a changeling as in so many tales of faery. He begins life as a smudged tiddler, then disguises himself in a smolting jacket of silver and pipe-smoke blue before slipping away by night, running the gauntlet of predators and disappearing into a marine life about which so little is yet known. From the once secret larders beneath the Greenland ice he then returns, with all the panache of a successful exile, arrayed in the gleaming mail of a knight-errant, mercurial, muscular, his body packed with fuel for the fasting ordeal ahead, which may last many months. At dusk, he glances across the tail of pools. He can travel 15 freshwater miles in a day. Shows as a blade at the falls – the river's hero.

As spawning time approaches, they swap silver for sackcloth. The hens grow sooty and gravid, the cocks turn to epigamic rust and develop a cartilaginous kype on their jaws for seizing rival suitors on the redds, where the hen with her tail aquablasts ridges in the gravel, laying nectarine-hued eggs that her mate, quivering and mouth agape, fertilises with his squirts of seminal milt (D. H. Lawrence became typically ecstatic about this process: 'Who is it ejects his sperm to the naked flood?' he asks in his breathy poem 'Fish'). Sexually precocious male parr – Nature's failsafe against dynastic disaster – often nip in and do their bit too, consorting with females some fifty times their size. And so the hydraulic saga continues, with its essential elements of love and death – for most Atlantics die after breeding once (and all Pacific salmon perish, their corpses enriching an immense food chain).

The spawned kelt is, in its own way, a remarkable iteration of this protean fish – a ghost of its former self, a ghastly, emaciated revenant. The 'long salmon', 'strike' or broken fish is a spent force of Nature, once pitchforked out for pig food, considered a mere irritant when one takes your lure in among the snowbroth currents of early spring and gives up easily with that distinctive, disappointing wobble. Personally, I admire them – poignant yet still heroic as they attempt to make it back to the salt, though now more scabbard than sword. You might also encounter an unspawned female – known as a 'baggot' ('bagged' signifying pregnant), 'thwarted matron', or 'rawner' (from the Nordic word for roe) – which would be taken for the table, and were regarded as 'gillie's fish'. These days, for conservation reasons, none of our salmon is killed in spring, irrespective of which direction they are heading.

In Britain, the kelt is an accidental (though sometimes welcome) bycatch for the hunter of increasingly rare fresh springers, but in April 2007 I joined a party on Canada's mighty Miramichi, where such 'black salmon' are deliberately targeted immediately the debacle of winter ice breaking up has occurred. This river system (about 100 miles of main trunk, plus several serious affluents, similar to Tweed) boasts the largest run of salmon in the western hemisphere, but they don't arrive until June; meanwhile, the kelts linger in river to intercept an upstream migration of smelt – *Osmerus mordax*, a cousin of our *O. eperlanus*, once very numerous, popular as a breakfast dish, and supposedly redolent of violets. The Canadian salmon are, therefore, most unusually, feeding actively in fresh water, putting on valuable weight for their perilous transit to safety across the Bay of Fundy, with its legions of ravenous seals.

The commencement of 'black salmon' season signals something of a free-for-all: it marks the lifting of the winter siege locking in the valley. All manner of precarious local canoes and watercraft are launched, and we did have to dodge the odd ice floe. As they become mended, the spent 'slinks' tend to strike your streamer fly with agreeable determination, and one chill day I boated a dozen, the longest almost at the magical 40-inch mark. It is not all catch and release, either (or *prise et remise*, as it is sometimes elegantly termed in the Maritimes): a state provision dating back to the times when fresh meat was scarce during these months allows you to keep some, but one grizzled guide, asked for his recipe, replied, 'Now boys, if youse cooks 'em just right, with 'erbs and spices in 'is belly, youse can gettem to taste almost as good as *shit*!'

We drank in Tom Thumb's favourite inn, were feasted by the Mi'kmaq First Nation tribe, and at the annual Ice Out ceremony (where seasonal fisticuffs occurred) I was mistaken for the prime minister of New Brunswick, since I was the only guest wearing a tie.

*

Our days at Mertoun came to an end in 2001, when my father's balance grew too unsteady even to be safe perched on the swivelling 'paddock stool' of a boat. Mind you, he finished his fishing career with a flourish, landing three while being roped down the Bridge Pool, and refusing to be pulled ashore to let his son have a turn. When you become really familiar with a stretch of water, its contours and moods enter your memory like a piece of music, and percolate your imagination. I never did fish with him again, and have not felt the inclination to cast my line any more on that storied beat.

The family connection with Tweed held on by another thread, however. My uncle had sold his fishings at Upper Hendersyde to the debonair young Sandy Gilmour, who in later life coincidentally became a friend of mine. Each November he was generous enough to invite me for a few days – staying at the Ednam, where his party trick was to toboggan down the stairs into the busy dining room on a tray – and we caught some tremendous fish. Against all the odds, I once landed three salmon over 20 pounds in one morning. Even when he became unwell and could no longer fish, Sandy would valiantly make it down to the river to urge on his team of guests; his ashes were at length scattered in the Mill Stream, and his jaunty hat floated away with them towards the sea.

Even then, we realised it was the beginning of the end of something. Tweed fish seem historically to follow cycles of roughly thirty years: it was an autumn river in late Victorian times, then gradually reverted to spring (the autumn nadir being during the Second World War) and now it seems as if the late-season fish have drastically fallen off again in numbers, as have Atlantic salmon in general, for multifactorial reasons. As the fisheries scientist Daniel Pauly has pointed out, each generation adjusts its concept of the normal along a series of 'shifting baselines' – where once we anticipated plenty, now a single fish a day (if that) seems reasonable. It probably didn't help that our faith in the Doctrine of Plenitude – an imperial assumption that the Almighty would never allow his natural resources to dry up – prompted us to hit so many fish on the head with our wooden priests. I have not had a wild Scottish salmon in my freezer now for quite a while.

Through the munificence of its proprietress, I still often fish Tweed at Dryburgh, upstream of Mertoun and certainly as scenic – 'round the noble sweep where the river holds Dryburgh lovingly in the crook of its arm', as the folklorist Andrew Lang described it. He savoured the trout fishing opportunities here, though he was far from adept as an angler. By the footbridge is a folly erected in memory of the poet James Thomson, another devoted fisher, who wrote 'Rule, Britannia'; his father was the local Presbyterian minister, and died when his head burst into flames during an exorcism. In Todholes pool here, John Buchan records how poachers once netted out an immense salmon which had been lost at dusk by Earl Haig, who hooked it up at Bemersyde and was later gleefully informed by the porter at St Boswells station (those were the days) that half of it, in a sack, had weighed

35 pounds, and it was therefore 'the biggest saumon that ever cam' oot o' the water o'Tweed'.

So, each autumn I still make that drive from Kelso towards Mertoun Bridge, Smailholm Tower standing like a warrior in the stubble fields with the Eildon pyramids beyond. These last few seasons we have hooked very little. They are gone, for now, that astonishing race of late-run fish, along with the lucky habitués who pursued them – and I sense neither will reappear during my lifetime. The hotel feels like some abandoned film set – the bar is almost empty, a spent force, a keltish version of its former glory. The hall floor is innocent of many corpses, but its mixture of polish and woodsmoke still transports me back to those heady, hurly-burly days, and I don't need any scallop-shaped madeleine dunked in lime-blossom tea to recapture such times gone by. These memories still surface easily, like a fin breaking the water midstream.

7

OF NOTHING

'I remember when I was a little boy, I felt a great
fish at the end of my line which I drew up almost to
the ground, but it dropt in, and the disappointment
vexeth me to this very day, and I believe it was the
type of all my future disappointments.'

Jonathan Swift, *Letter* from Dublin to
Alexander Pope (5 April 1729)

The 2nd Earl of Rochester was a contemporary of honest Izaak
Walton, but this rakehell libertine courtier would have had
little time for fishing. He was a known intimate of citrus Nell –
who certainly angled with Charles II – so it may be he sought
the monarch's approval by accompanying them, though he
was quite busy with other country matters, and even by

the standards of the Restoration court he was gloriously debauched, said to have been uninterruptedly drunk for five years. As a teenager he was sent to the Tower for kidnapping an heiress, then set himself up as a quack fertility consultant. He once destroyed a rare royal sundial, with the remark, 'Doest thou stand here to fuck time?' This stylish iconoclast died aged thirty-three, literally clapped out.

His poetical accomplishments included a hymn to the Dildo, but also an astute philosophical composition ('Upon Nothing', 1679) addressing the concept that since God created His world from nothing, then the great Negative must be the guiding principle of all Life – the poem ends with a kind of swirling black hole in which everything revolves around nought: 'The great man's gratitude to his best friend,/ King's promises, whore's vows – towards thee they bend,/ Flow swiftly into thee, and in thee ever end.' Some astrophysicists aver that the only three numbers in the universe that make sense are zero, one and infinity. With angling, zero is our fundamental condition – it's where you always have to start, and it looms all day as where you may also be condemned to finish. I'm not sure this applies to other pastimes, such as trainspotting or scrimshaw.

Failure is the necessary yardstick of our pastime, and the inevitable blanks and 'skunks' comprise one of its several tribulations (though judging by the accounts of certain angling writers, setting forth arrayed like SAS troopers who have forgotten to shave, success seems the usual outcome). Being a gloomier protagonist, my maxim is the Melanesian pidgin phrase, *Olgeto taim mi go huk, me no painim me no painim waupele pis* – whenever I go fishing I never catch a fish. If *nada* is your baseline philosophy, some expeditions will agreeably surprise you.

One question that baffles civilians is, 'How can you still enjoy a day if you catch nothing?' I don't think Lear was giving us the full picture when he growled at his daughter, 'Nothing will come of nothing.' A fishless day does not have to be a joyless one, but never believe someone who professes not to care if they catch or not ('It's just fun being on the water'). Yes, it is wonderful to be on the water, but every angler prefers to hook a fish. What I think we mean is that we would prefer to be fishing than doing anything else (though I imagine the alternative temptations of, say, an afternoon in some Cuban seraglio with a pipeful of golden hashish might give a few anglers pause before opting for their 8-weight rod).

Are there silver linings to be sought, if your day results in zilch, or the Big Egg? It rather depends on how you calibrate angling success. For me, it's less about setting goals, and more to do with managing expectations, though the competition fisherman and specimen hunter might not concur. You can have fun trying, and derive some satisfaction from knowing you have fished well (after all, no one brings back much from a niftily played round of golf): fly-casting, in particular, seems to offer some aesthetic satisfaction, properly performed. There is a tradition of extolling the compensatory delights of the sport, dating at least as far back as Dame Juliana, and nicely echoed by Robert Burton: 'And if so be the angler catch no

Fish, yet he hath a wholesome walke to the Brooke side, pleasant shade, by the sweet silver streames ...' (*The Anatomy of
Melancholy*, 1621) – though I don't recommend trying that line
on your empty-handed mate when you are three fish ahead
and celebrating in a luncheon hut plastered with trophy photographs like the waiting room of an assisted conception unit.

A lot of angling runs along lines of serendipity, and your
'clean' day is effectively the flipside. Sporting etiquette dic-
tates that you should take unsuccess on the chin, *à l'impériale*,
but that can prove tough. There are times when no amount
of brave-faced exclamations ('two more and I'll have a brace!')
can mask the fact that a blank can seem like Doomsday.
Nobody has better examined this taxonomy of grief than
that ever-astute commentator H. T. Sheringham, in his essay,
'Blanks and All About Them' (1916), in which he analyses
three different categories – the Blanks Positive, Comparative
and Superlative. I can think of several rankling examples, each
one still vexing me to this very day.

The Blank Positive describes a simple state of fishlessness.
The frequency with which these occur may depend on the
type of angling you favour: for instance, the pursuit of large
carp, or subtropical permit, is often articulated by days when
one's reel is never turned, whereas chasing stocked rainbow trout in the smallish confines of a put-and-take pond
might offer a strike rate more akin to a spell of falconry in
Trafalgar Square.

For thirty years now I have adored fishing the saltwater
flats, not least because the species variety there usually means

you stand a good chance of catching some creature or other, provided one is catholic in one's tastes. The Jardines de la Reina archipelago off the south coast of Cuba is somewhere I have visited eight times, and in the early years of its being opened up to foreign anglers (though strictly not to the *yanqui*) its mangrove creeks and turtlegrass expanses were so under-explored and abundant in marine life that my old boat partner 'Don Jon' once said they made the Florida Keys look like the shelves of a Soviet supermarket (this was back in the 1990s, I should stress). Some days we would catch twenty bonefish to our boat, often with several snappers or jacks as a bonus. Even when the weather was squally and visibility poor, there was usually somewhere we could score a fix of sorts.

One afternoon hot as the hinges of hell's gates, we entered a bay we had never visited before, and it appeared that the entire area was being cut by dorsal fins and tails of feeding bones, occasionally quivering like banners as the fish tipped up and snacked on the marl. Jonathan had already taken a few fish elsewhere that day, though I had not; when we beheld this prospect – an idyllically target-rich environment – I actually whooped aloud. Over the next two hours, my companion began steadily to pick off the bonedogs, while I, using apparently identical flies and tactics, could not manage a single lousy take. I felt as doom struck as the author of that Yiddish saying, 'If I dealt in candles, the sun would never set.' I became progressively frantic, and got into a mental rut of the type to which I am idiotically prone (especially around the middle of a week's trip when frustration may have become compounded by biting flies, heat so strong you can practically hear it, too much smoking, or sundry other factors). I call this the Bonefish Blues. What happens is that I lose all confidence, begin casting like a gibbon, feel the situation has become undefuckable, and wax defeatist. This nihilist mindset is fatal to success: you have to fish your way out of it (*Solvitur pescando* is the only motto). It's a bit like that with writer's block, which is certainly no joke: you just have to scribble your way out of it. As Abe Lincoln liked to observe, 'Every man got to skin his own skunk.'

On this occasion, I felt like a wallflower at an orgy. Being a diplomatic sort of partner, Don Jon settled down on the cooler box, ignited a powerful Cuban candle, and let me continue until I had ultrasonically shattered my jinx by sheer force of repeated casting into the seething bouillabaisse of *Albula vulpes*. Eventually, there came a take even I could not foul up, but when it was hustled to the boat the fish proved to be an undesirable, minuscule barracuda. Imperilling an ancient friendship, Don Jon beamed through his rich folds of Maduro and said, 'I reckon it takes almost preternatural powers of selectivity to locate the only baby barracuda in this entire area, and proceed to hook it.' At moments like this, you just have to acknowledge that the world sucks rocks, and you have undergone a Blank Positive. We skiffed back to the mothership, in search of rum service.

There then comes the Blank Comparative, which involves the disparity between one's high expectations, or the reputation of that venue, and what actually results (it may not comprise an absolute skunk, but there will be woeful out-comes). One August in the 1980s I was a guest with my father-in-law (the Doctor) on the fabled Hebridean loch system at Grimersta, when precisely that befell us.

I already knew the place a little, having fished there with my uncle and several elderly syndicate members, and indeed it was there I had landed my hundredth ever salmon, which resulted in having to buy champagne for everyone in the Lodge. It was also here that, on 28 August 1888, an artificial spate having been created by releasing impounded water following a severe summer drought, Mr A. M. Naylor, a land agent from England, set the British record for numbers of salmon landed by an individual in one day: fifty-four to his own rod, for a total weight of 314 pounds (the next year he gave up fishing altogether and took up big game hunting). A famous black-and-white photograph shows three densely bearded gillies in deerstalkers displaying his catch slung on poles. These were the waters for which Arthur Ransome developed his Elver fly, a fine example of which is framed over the Lodge fireplace – so it is a fishery one might well call 'storied'.

The Doctor and I had drawn Beat Four, towards the highest part of the system. On the way up, we fished the Ford Stream, where he took two fresh fish and I had one. On the first drift of the loch itself, off The Weeds, I saw a nice head-and-tail rise to my dropper, though the fish never touched it. 'We shall have a good day, gentlemen,' our young, mutton-chopped oarsman announced confidently. They say: If you want to make God laugh, tell Her your plans.

We never saw another living fish all day. When we arrived back at the Lodge, it turned out the other three beats had taken sixty-five between them. One angler had killed thirty on his own. In fact, all the salmon were slain, and laid out on the grass together. I have never seen a comparable sight, in angling. 'Happily, however, the pleasure of fishing is not measurable in weight of fish,' admonishes Ransome. As a salutary reminder of the vagaries of fortune, I have pinned to my tackle-room wall a photograph of those sixty-eight salmon, with a circle around the solitary grilse in the bottom right-hand corner, stating, 'This one was mine.'

Sheringham attaches the Blank Superlative to expeditions characterised by monumental, heroic failures and missed opportunities. The day I have balefully in mind added to my personal mishaps the blithe success of two other companions, and so I propose renaming this category the Blank Pancatastrophic, since we all secretly know that the happiness of others only compounds our own discomfiture (as Gore Vidal observed, 'Whenever a friend succeeds, a little something in me dies'). I am always surprised at the level of temporary loathing – that acidic seep of childish envy – I still experience when others do well with rod and line, wishing upon them a sudden attack of the bloody flux, or testicular perforation when next they cross barbed wire (and these are my friends and family). So it goes.

We were on the Ponoi in Russia's Kola Peninsula one late August day, and guide Rick sounded optimistic as he beached the boat at Hard Curve for fellow Brits Alistair, Robert and myself. The river was in apple pie order, and some of the first fish of the autumn run had apparently arrived on the previous night's high tide. I all but whipped out the abacus and began counting my chickens. After all, I was a guest on one of the world's finest salmon streams, developed for Western access only a few years previously, and I'd taken a few fish already that week, including several on a dry fly. I might even have allowed myself a slight swagger as I headed towards the neck of my first pool.

I proceeded to fish as poorly as I had for years. I began striking at offers, letting fish have slack, breaking off on the back cast (this was before I learned the Spey Throw, and its various iterations), and even contriving to tear a hole in a wader leg. I might as well have been fishing with rubber hooks. When we met for our shore lunch of broiled salmon – one my colleagues had killed, having released a further seven – the others all murmured their commiserations, though in truth it's no mystery how we always summon the strength to endure the misfortune of others. At the end of that day I was comprehensively waterlicked. I had succeeded in properly hooking one salmon, which found a boulder whereupon, to use that genteel Edwardian phrase, 'I was incontinently snapped off'. Less politely: I felt that fuckpig of a Russian river was shitting all over me – as John Rochester might have put it (though that particular noun of endearment was apparently coined by Lawrence of Arabia).

Back at the Ryabaga Camp, there was a hullabaloo of celebration. Jack, a Louisianan judge, waved his baseball cap at me as we disembarked and cried, 'Hell, I guess *everyone* caught a buncha fish today!' I fixed my politest smile like a bayonet, and refrained from directing him to shove it up the tooth fairy's backside. I was the only rod in the place who had blanked. I felt as solitary as a bastard on Father's Day, as they say Down Under. I was as bewildered and out of place as Schrödinger's dog. In the gloaming, shaking my unslimed fist at the infernal mechanics of the universe, the dark and loathsome wings of the world crimped around my hunched shoulders, I slunk alone towards the *banya*, clutching the remains of my duty-free Laphroaig, and sank sulkily into its chemical embrace. Right then, I did not want to be consoled: I wanted to sweat the misery out of my wretched body like so much mozzie toxin. Then – half-man, half-Scotch – I slouched back to my tent, under the bitter stars.

After a night of ragged sleep I awoke with a dragonish thirst, and made it to breakfast looking like Death in a White Sauce; but later that day, down at Gold Creek, guide Dima put his net under ten beautiful salmon for me, and the world was once again a realm of glory.

In 1991, I made a round trip of 16,000 miles for a Blank Comparative. We were a party of three, and attracted a degree of attention at check-in since my fellow anglers were a building contractor from Sussex and his inamorata who resembled 'Hot Lips' Houlihan from *M*A*S*H*, but clad in riding boots and an ankle-length mink coat, and we were the only civilians about to board an RAF Tristar to the Falklands, not the most popular of unmarried postings just before Christmas, a fact perhaps reflected in the cranberry-eyed, shaving-nicked demeanour of the rest of the breathing freight. Forewarned the eighteen-hour flight was dry, I had emptied an opaque diarrhoea medicine bottle and filled it with vodka. Even when we touched down on Ascension Island – once deemed a ship of the line, so any children born there were registered at Wapping – we were forbidden beer. On arrival at Mount Pleasant (nicknamed the 'Death Star', and boasting the world's longest subterranean corridor), we were decanted into a briefing room. Outside, the rain was coming down like buckshot on the scarred and streaming landscape. It was almost the austral midsummer.

Never exactly billed as a honeymoon destination, the Falklands were described by Darwin as 'an undulating land, with a desolate and wretched aspect', while Louis-Antoine de Bougainville (after whom a fancy vine was named that is a favourite of travel writers looking to lyricise their prose) visiting in 1764 noted 'a vast silence broken only by the occasional cry of a sea monster'. This is not ideal Tourist Board copy. The treeless terrain is indeed severe – like the Isle of Lewis, but instead of sparse heather there was moorland diddle-dee, scurvy grass and the odd stand of tussac. Great 'stone rivers' of grey quartzite cascade down some of the hills, a topography familiar from images of the 1982 'Conflict' which was provoked by an Argentinian scrap-metal dealer landing uninvited on South Georgia. Apart from the ensuing warfare – which involved thirteen thousand invading enemy troops armed with stocks of napalm – the islands have remained relatively free of violence and crime, though one senior cleric whispered to me that the 'kelpers' (as residents are known) had an occasional inclination towards incest.

Initially we stayed at Port Howard as guests of Robin Lee, a reserved but cosmopolitan kelper whose family had been ranching 200,000 acres there since 1890. The islands grow sheep for wool (Robin ran 42,000 pure-bred Corriedales) but at that time European regulations forbade the importation of Falklands mutton, or even supplying it to the military, although the only recorded instance of food poisoning had come from an army chef microwaving regulation meat pies.

It was tempting to quip that when it's 09.30 in Greenwich it's still 1953 in Port Howard, except that everywhere were reminders of the fighting. Some bogs were still sown with mines, and we came across the fuselage of an Argentine Mirage jet with overpainted Israeli markings. We saw the spot where SAS Captain John Hamilton MC was shot in the back while covering the escape of his signaller. There had been over a thousand Argentinian troops stationed at Port Howard, and one of their officers had spat on the Queen's portrait hanging in Robin's office, ordering him at gunpoint to take it down. He refused.

We were here to fish the River Warrah for its multitudinous sea-run brown trout. The day I'm remembering (23 November) begins with a 10-mile Land Rover drive across open terrain, *The Archers* on the radio. There are no roads, though every time he approaches a ditch Robin looks ostentatiously both ways and says, 'All right. Nothing coming.' It's no country for the faint-hearted, and is not even that other travel-writers' cliché 'a land of vivid contrasts', despite an occasional slash of gorse across the bleakish panoramas. We scatter flocks of Upland geese, and the odd Paraguayan snipe. The estate has a thousand gates, and we use six of them – an agreeable ritual requires you to sup a dram as you close each one. It's a far cry from Ambridge.

When we arrive at the Green Hills tributary, the wind is 'blowing a bastard'. On a patch of bog balsam, a turkey buzzard is snacking off a sheep's eyeballs, like some football supporter scoffing pickled eggs in a post-match bar. A light blizzard has begun to illustrate the slopes – the weatherism that 'you can get all four seasons in one day' applies for sure in the Falkland Islands. But finally I get to tackle up my travel rod with its 7-weight sink-tip line, and show Robin my fly box: even down here, practically at the end of the world, our chosen fly pattern is a trusty old Teal, Blue and Silver.

The river is running high and peaty, gargling around cut-banks and describing plenty of foam lines. For three hours without respite I fling my fly into the squalling air, never less than a Beaufort scale 7, but when we stop for a traditional 'smoko' break I have moved, touched and seen precisely diddle-dee squat. Not even the fin of a finnock. 'Of course, the fishing can be very good at times,' remarks Robin levelly, 'but it can also be bloody awful. Still, in thirty years I've never got this far downriver without coming across a fish. Never.' God grant me strength. As Mark Twain reputedly remarked,

'There's no use in your walking five miles to fish when you can depend on being just as unsuccessful near home.'

Thanks to the recent eruption of Mount Hudson in Chile, the islands were apparently coated in a volcanic ash so abrasive it was wearing down the teeth of the livestock, and furthermore it had leached into the headwaters of all the Falklands rivers, flushing almost every migratory fish back out into the safety of the sea. All had come to naught. Over the course of a longish fortnight, we managed the odd one by hurling great Toby lures into the fjords, and I discovered some bronzy-silver linings in the shape of obliging estuarial mullet (which were about as tasty as a Miramichi kelt) but that was not what I had come for. Still, to use Robert Burton's phrase, I suppose I had enjoyed 'a wholesome walke to the Brooke side' – if nothing else.

Just a few months later, on 25 March 1992, Hebridean guide Alison Faulkner landed the Falklands record sea trout, weighing 22½ pounds. The fly was a Teal, Blue and Silver.

A refined state of Nothing is the void where Something has been. In 1911, tourists flocked to the Louvre to stare at the space on its walls where the stolen *Mona Lisa* had formerly been hanging. And of all the cartoonish stereotypes familiar to the layman about anglers (along with the hooking of discarded boots, and the telling of tall tales) is the notion of The One That Got Away. This fuels the sob stories of a sport elaborately spattered with spilt milk, but in practice there's seldom much amusing for the protagonist about that pang of severance, the nauseating instant of twangling discontinuity which is often accompanied by vernacular oaths that would have had Captain Haddock flushing to the roots of his beard. The mildest such retort I have come across is when the serf Tuman (in Turgenev's *A Sportsman's Notebook*, 1852) loses a nice perch and merely exclaims, 'Oh, he's got away, the Chinaman.'

There's also a Chinese proverb which runs, 'The whip that's lost always had a golden handle', and it's true that a significant fish which escapes has a tendency to grow in the memory, though fish can seem exceptional without being unusually large. Things are worse when you have had a sight of it, or the quarry is very nearly 'yours', and then it can be no small matter. Ransome reckoned the sensation probably resembled Lucifer's feelings when 'first dumped in Hell' – an amalgam of impotence, incredulity, shock, post-coital *tristesse*, and burning indignation. Anticlimax doesn't nearly convey the engulfing enormity of this hideous divorce. I believe this experience seems to matter so passionately because the act of fishing temporarily draws a line (which I call 'the lightning thread') between you and a shadow world, and it's eerily disconcerting when one of its denizens fades away from you again, along with your highest hopes, like Eurydice being lost to Orpheus as he had almost spirited her back from the Underworld, drawing her from darkness into light – the essential process of angling itself. Indeed, the fish became a symbol of the Greek cult of Orphic worship (and if you think this is all histrionic claptrap, feel free to go and buy a niblick and putter).

Maybe the most succinct evocation of this lingering afterlife of lost fish comes from Norman Maclean, reflecting on the departure of a hefty Montana trout in terms both demotic and Wordsworthian: 'Poets talk about "spots of time", but it is really fishermen who experience eternity compressed into a moment. No one can tell what a spot of time is until suddenly the whole world is a fish, and the fish is gone. I shall remember that son of a bitch forever.'

Walton himself, whose Chapter XVI 'Is of Nothing', and concerns the pleasures of music, as opposed to angling, has Piscator firmly correct Venator thus, 'Nay, the Trout is not lost; for pray take notice, no man can lose what he never had,' which is true, but ultimately unhelpful. We know what we mean by loss.

Each angler will have his or her catalogue of personal woes. I imagine I can still see that brownie from New Zealand's South Island which I hooked on a green beetle pattern, and before I could put the heat on him ran my leader around a willow root. I once missed twenty-two baby tarpon on the trot, through inexperience; and three grilse came unbuttoned in ten July minutes from the Spout pool on Arndilly – sometimes you have to take the rough with the rough. That Restigouche salmon still surges through my dreams, as if on some brief loop of appalling repetition – in Cicero's phrase, *'abiit, excessit, evasit, erupit'* (you get the drift). Playing a modest-sized, fly-hooked tiger-fish on the Zambezi, I watched as a by-God hog of a grandpappy tiger lunged up out of a bankside hole, seized it in his jaws, and if only I had just … ah, but there we have it: as a Bahamian guide once reminded me, 'If the Queen had balls, she be the King!'

The ever-sagacious Sheringham, in his essay 'Hooked and Lost', tackled another issue: is it better to have hooked and lost, than never to have hooked at all? As a rookie angler I would have answered in the affirmative, but now that the majority of my seasons must be behind me I'm inclined to agree with the essayist that it might (for the remaining shreds of one's sanity) be better not to have had so many regretful encounters, wherein that momentary hotline to the universe is cruelly disconnected. Besides, I think his analogy is misleading: it more closely resembles an act of unrequited love, as there has never truly been possession. For a genuine example of the 'loved and lost' syndrome, surely nothing could rival the emotional complex of Hemingway's heroic Santiago (in *The Old Man and the Sea*, 1952) who endured eighty-four blank days, only to hook a 'grander marlin', which sharks systematically tore to pieces from the side of his boat, leaving just a skeleton that measured 18 feet. That is one that was genuinely both caught, and got away, and subtly describes the phantom limb syndrome of our own, lesser experiences.

Even with the publishing triumph of *Gulliver's Travels* recently behind him, Dean Swift, a man morbidly disposed towards dejection, wrote that letter (which is the epigraph above) when he had failed to land the bishopric he felt was his due, and which he took as a snub for the rest of his days. It is no casual remark in which, now in his fifties, he recalls the intensity of childhood disappointment, an incident that occurred on the River Nore when he was a schoolboy at Kilkenny. I don't suppose anyone dared console the young misanthrope with quips about Long Distance Release.

One of the classic war stories about salmon concerns an actual bishop – the Revd G. F. Browne, later Bishop of Stepney, then Bristol, who was trailing his blue Phantom minnow from a boat rowed by his gillie Jimmy at the confluence of the Tay and Earn one October noon in 1868, when he was taken by a fish 'as big as a well-grown boy'. Fortified with potted grouse sandwiches, cakes, cheese and whisky which were ferried out to him, our divine fought his 'Tartar' through the gloaming and into the autumn dark until ten-thirty that night. Strand by strand the line began to fray, but on moral grounds he refused Jimmy's suggested ploy of luring their adversary to the surface with lantern light, and thus securing it on the gaff. The fish suddenly began to flail as the tide changed; it ran towards the boat, and then, 'Up comes the minnow minus the tail hook. Jimmy rows home without a word, and neither he nor the fisherman will ever get over it.' A possibly apocryphal report relates how, the following year, a fish weighing 74 pounds was taken in the nets at Newburgh, with the scar marks (stigmata?) of an identical triangular hook in its jaws. Either way, Browne lived until the age of ninety-six, and I wager he frequently rehearsed the details of that loss, despite the consolations of his strong faith. The fish would have been our British rod-caught record.

This episcopal theme recurred more recently, in 1989, when

on Alaska's Kenai river a Minnesotan angler, Bob Ploeger, was plug-casting for chinook salmon with his guide Dan Bishop. They latched into a magnum unit which looked clearly to be over the record, magical 100-pound barrier for a King. He fought this lunker from the powerboat for thirty-seven hours, the contest being covered by television crews, and food being provided by the local hamburger joint. Eventually, the gears on his reel failed, and Bishop decided to have a go with the landing net anyway – whereupon the hooks snagged and were pulled out of the fish's mouth, and it escaped. 'I was never good at endings,' sighed Bishop (a perilous endgame indeed). Displaying remarkable sangfroid, Ploeger took a brief rest, described himself as 'A little disappointed, actually. Not discouraged', and resumed casting. Jesus God.

Things are considerably worse when the escape of the fish is a patent result of your own mismanagement. Pilot error, shoddy knots, sloppily loaded lines or laziness in prepping gear leave you with a pang of merited disgrace – a lurch of self-disgust to add to the rest of your swirling problems. When you know the fault was yours, there is no drawing up a counterpane of resigned fatalism to warm away that bone-chill of failure. Neither age nor distance will soothe such a memory of one lifetime's opportunity botched for all eternity. Zane Grey truly wrote, in his opening to *Tales of Fishes* (1919), 'To capture the fish is not all of the fishing. Yet there are circumstances which make this philosophy hard to accept.' I'll buy that for a dollar.

Not many months before he died, I was driving the

redoubtable Fred Buller home from a day together on the Test at Mottisfont (F. M. Halford's old beat), when for the first time he began talking to me about the historic loss of his gigantic Loch Lomond pike in 1967, the result (as he had meticulously described in print) of an inexcusable knot failure. Fred was beyond doubt the greatest luciophile of our times, and this was the ultimate trophy he had been hunting, for decades. The magisterial Richard Walker (no stranger to record fish) saw it swim past his own boat, and put it at over 50 pounds, adding, 'It frightened me.' Fred had never expressed any regrets to me before, and had caught so many remarkable fish during his distinguished career that I never imagined the scale of this particular affliction, but that evening he laid his hand on the dashboard, and simply said, 'You know, it's been a nightmare to me ever since.'

Off Saint-François atoll in the Seychelles, I had my billfish comeuppance. I have indulged in a certain amount of *pêche à gros* (once, off Mauritius, I lost a sizeable blue marlin when, after forty minutes, it shredded my nylon leader like a lawn strimmer) but on this occasion we were trying for sailfish on the fly, and I was hopeful of being awarded a coveted 'Bills and Bones' badge by the lodge, as that morning I had already taken twenty bonefish off the flats, plus our only giant trevally of the week (though it was bitten in half by a reef shark before I could bring it to hand, and all I reeled in was the cookie-cut head section). As the skipper rigged up my 12-weight, I suppose I was dangerously suffused with hubris.

The technique is to raise a 'sail' behind the launch by

trolling a hookless lure to coax it up towards the prop wash, then whip away the 'teaser' and let the angler drop his Pink Popper tandem fly and chug it in front of the fish as it casts hungrily around for its prey. Talk about fin fever! Barely ten minutes after the start, a sail appeared slashing at the plug in a state of great excitement; I lobbed it my offering, it arced gracefully across the surface, turned away with its prize, and I strip-struck, as instructed, zealously. With a retrospective shudder I can rerun those oh-no seconds when I realised that in the process I had looped the fly-line around the butt of my rod. The skipper spotted it too, and together we rushed for the stern – but there's no reclaiming 6 inches of taut line from a billfish that, in a violent fit of buyer's remorse, is rocketing directly away from you so rapidly that 'the flames are coming outta his ass'. A carwash of adrenaline sprayed through my system, as I saw the rod explode with the sudden crack of a pistol, and the line parted.

Something heavy sagged within my viscera. I felt as gutted as a barracuda's breakfast. From flash to bang it had perhaps lasted two minutes. The Chinaman.

One thing I do know, buddy: It's a Long Walk Back from Shit Creek.

8

HINTERLAND

'That memorable few hours in your paradise. I keep
recalling it. That intense Prospero's cove, the sea
coming in so courteously, so polite and diffident,
right to your feet.'

Ted Hughes, from a letter to my parents-in-law,
Elaine and Alasdair Fraser (10 September 1991)

I had a pretty good inkling the Doctor disapproved of his
lovely daughter's new boyfriend by the way he disappeared
into the pelting dark, along with the only torch. I was a
long-haired, bumptious undergraduate toting a brace of inap-
propriate Asprey suitcases and a carton of dead game from
my uncle's grouse moor, and as I teetered along the 2-mile
clifftop track, the moderate storm dashing my luggage against

my legs, I eventually beheld below me a small, lit dwelling which I took to be the gate lodge to the Fraser family estate. It proved to be the Doctor's beloved croft house – and none of us would have guessed that we would still be visiting it forty-five years later, nor that Helen and I would eventually become engaged there, and our son James in his turn, and that it was to prove a place of such happiness and respite. The Doctor was heard to refer to me as a 'Hello Henry' (maybe it was my Gucci slippers), but the day came when we became inseparable angling companions.

The house is named Molban, for the beach of pale stones where it stands against the sea. That first morning I awoke to the sound of the tide slobbering into rock pools and sucking on pebbles just yards from the front porch. Across the stone-washed sky squadrons of gannets in their sulphur skull-caps were heading for the feeding grounds as the Minch boiled coldly beneath the headland, throwing up shards of gunmetal light. Along the tangled strandline, starlings strutted, cocking their heads like invigilators. It certainly did not feel like home.

At the top of the brae behind the house is a long and lovely loch. An unfiltered hosepipe conducts its hazel-tinted water directly down to the taps, making a soft cup of tea and correcting your evening dram when, in Keats's nice phrase, the water is 'diluted with a gill of whisky'. I tried that loch before breakfast the first day, and returned with two brownies – to the further chagrin of my future father-in-law, since it has a reputation as a dour and challenging water. Over the decades, I have probably fished it more than any other place, and have come to appreciate its moods (it has effectively attained the status of Home Water) and read its contours – the cranny beneath those stunted and salt-burned rowans, the spot where a south-westerly channels food lanes through the narrows: sometimes it has the dull look of abandoned soup, but occasionally it scintillates and obliges. I felt I was off to a good start.

Trailing like a kite off the north-west coast of Scotland, the Outer Hebrides is an archipelago comprising several hundred islands named in the Gaelic 'The Land of Strangers', while others know it as God's Necklace (Hebrides itself appears to be another ghost word, a scribal error dating back to Pliny). The Long Island is a world apart from the tartan and souvenir spurtle image of the Highlands – the inhabitants say, 'On a clear day, you can see Scotland.' A census just before my first visit recorded some four hundred monoglot Gaels. The proud insularity is subtle and wry – one ninety-year-old man I met was nicknamed 'The Yank', because in his youth he had once been to America. Island time can also bemuse the newcomer: discussing the Mexican concept of *mañana*, one Harris native replied, 'No, I don't think we have any word that expresses quite that sense of urgency.'

Much of Harris is Archaean gneiss – great dove-grey mammillations of rock that sometimes glint with mica and, at three thousand million years old, are among the most ancient parts of Europe. Up close, these lithic faces appear dynamic, swirled with omega or ampersand shapes, Munch-like contortions with thin mapworks of lichen. On the higher slopes, scree lines give way to boulder fields of glacial erratics and it is widely referred to as a lunar landscape. Some of this rocky skeleton is wrapped in a blanket of peat, its dark rind exposed and leaking where the hags have been worked for winter fuel. But Harris presents two Janus aspects. On its west coast, there are dunes of blown shell sand threaded with marram grass like hairs on some old man's scalp, and this has enriched the machair grazings which, in early summer, can be a mass of sea pink, vetch and orchids. But on the east – particularly in the Bays area, where we stay – there are no such palomino beachscapes; the prospect comprises shattered headlands fanged with dark rock and skerries scarfed in bladderwrack. There was not even enough topsoil to bury the dead, and on the unmade track towards the cemetery at Luskentyre on the softer coast you can still see the stones where coffins were rested as their bearers changed position.

Gaelic is hereabouts held to have been the language of Eden, and you could regard it as a Paradise of sorts, but in reality there is precious little about the hardness of island life that could be called idyllic, despite generations of incomers ('white settlers') sentimentalising it. The oceanic climate is protean and fickle: as our friend Flora, born in Molban, liked to say, 'Oh, the North Wind here just blows from every direction.' I'm never quite sure about psychogeography, but it seems to me that Harris, with its immemorially severe demeanour and its hard-won beauty, might be a country of the mind.

Victorian visitors to St Kilda, an island group 57 miles even further into the Atlantic, imagined the place was an Arcadia, since they could only ever travel there in extremely clement weather. But these places were true edgelands, where a tenuous community survived until 1930. Though surrounded by fertile seas, the Hiortaich seldom fished for food, and even then discarded all but the livers. Angling from the rocks was forbidden, as it disturbed the sea-birds that for centuries were the staple of island life – gannets' necks were used for shoes, coral-coloured oystercatcher bills made pins, and breakfast was boiled fulmar with porridge. Like most islanders, they deliberately did not learn how to swim (it merely prolonged your struggle before drowning), and it was said no male St Kildan ever died in his bed: most of them fell from the crags while fowling, 'he went over it' being the phrase. I have stood inside the small graveyard that contains most of the islanders in history. At last, the remaining few – some of them not speaking any English – were evacuated to the mainland 100 miles away, having first drowned their sheepdogs and left in every abandoned house an open Bible and a handful of oatmeal, as a benison. Never having seen a tree in their lives, many of them were set to work for the Forestry Commission.

The dresser in the Doctor's croft is stored with sea returns – glass floats, dried mermaids' purses, the skulls of kittiwake and otter, whale vertebrae, a cantle of ambergris, even a rare St Kilda mail boat. Beachcombing is an essential part of island life, especially for driftwood – precious indeed in this place where trees are scarce. That obstreperous Victorian naturalist Charles Peel (who once bagged a grey seal, a curlew and a great northern diver in the same day) recorded Hebridean flotsam that included coconuts, a turtle, cases of champagne, a 60-pound cheese and 'tins of vaseline much appreciated by the natives, who spread it in a thick layer on their bread and ate it with great relish'. The most celebrated salvage haul was those thousands of 'Polly bottles' of whisky illegally rescued from the SS *Politician* when she foundered off Eriskay in 1941. Nineteen islanders were sent to prison, a detail that did not feature in the romanticised *Whisky Galore*.

For many years, Molban was kept in trim by a remarkable neighbour named Rachel Macleod – a gentle, devout lady in a housecoat and bonnet, who had lived alone all her adult life on the Point, and over decades kept a daily record of the weather conditions in her diary. Her grandfather's brother was known as The Bard (the only thing he feared was thunder, as, aboard a fishing boat, lightning had once sheared off half his moustache) and she herself had five generations of Harris history in her head, as well as much island lore. Like some Icelanders, she often spoke on the in-breath. Though she herself would never mention it, Rachel was considered to have second sight (the Two Sights is regarded as a grave responsibility, even an affliction, in Gaelic culture). She once told me of hearing powerful voices and seeing arcs of indoor lightning when she entered a nearby croft where, it transpired, a woman had long ago burned to death when her apron caught fire.

Although there is some high-end sport to be had in the islands, most of our spare time is spent rough fishing. The daily ritual of going to sea to catch bait for the lobster pots – drifting and jigging for mackerel and pollack, and the odd bonus cod with its profile of a pot-bellied alderman drooling Brown Windsor soup – precedes the lifting of the traps, and my inveterate loathing of crabs means I leave handling the ghastly, Spam-coloured partans to others. I am better disposed towards the lobster, with its greenish blood and inky cladding, those fluttering spinnerets – the Swiss Army knife of crustacea. When not confined in Dr Blanche's asylum, the deranged bohemian poet Gérard de Nerval is said to have paraded his pet lobster, Thibault (rescued from some La Rochelle fishermen), on a blue ribbon in the gardens of the Palais-Royal, claiming it was peaceful, did not bark and knew the secrets of the ocean. No wonder he was later lionised by the Surrealists. In 1855, he hanged himself from the grating in a Parisian backstreet using an apron-string he was convinced was a garter once belonging to the Queen of Sheba.

From the Doctor's venerable fishing boat we have harvested shoaling whitebait with nets improvised from nylon tights, followed basking sharks, spotted a submarine's conning tower, and been escorted by porpoises through the phosphorescence of our evening wake until the sun melts into the hills and the world becomes fire and glass. Once, adrift in a wobbly black inflatable dinghy spin-casting for cuddies (young coalfish), I heard a breathy exhalation behind me and saw an approaching killer whale in search of its mid-morning seal snack: I only stopped rowing my pinniped-resembling craft to safety when I was several feet up the beach.

Though many lochs now are extinct – unprotected, and poached out by netting – there used to be plenty of waters where you could go for a day after sea trout. The first week of my stay, Alasdair took me to the Sheep Loch, which he claimed was one of the finest such places. He insisted on rowing me, and when I duly landed a fish in excess of 4 pounds (it was on an unconventional Spuddler), I thought he was going to bite through the stem of his briar, as in fact he had just been trying to put me, the young arriviste, in my place – nobody had caught anything like that off the loch for several years, so my fluke merely incurred additional disfavour. More often we went in search of brownies, in lochans frosted with water lilies, usually yielding nothing larger than 'breakfast fish', but once in a few years you might see a trout with a head the size of a coconut, as you dabbled your dropper or a dry Daddy trailed downwind of the floss streeling from your dapping rod. You usually have to work hard for an island fish.

On the switchback Golden Road to Tarbert (the island's 'capital') the little townships are a jumble of traditional dwellings hunched against the elements, backyards clogged with defeated, discarded machinery, interspersed with nicely insulated modern kit houses sporting gardens of pampas grass and snapdragon. Scraps of black plastic twirl on the barbed wire like gibbeted witches. The land may look threadbare, scuffed and harried by the wind, but it's too defiant to be forlorn, and there is a curious lustre to its asperity. Tarbert itself sports the time-warp, family-run Harris Hotel, and in those early days the Doctor and I frequently took a day's fishing on their Lacasdale loch system. The uppermost one was our favourite, beneath the sentinel hills that lead to Rhenigidale, the last village in Britain not to be connected to a road (we used to see the postie trudging overland with his pack of mails). A gillie was needed here, to sidestep the boat through the pleated water along the shoreline towards the far beach – Norman the weaver, or Duncan, in his pork-pie hat, would net us the odd finnock and, twice only, a torpedo-shaped island salmon.

It was after just such a foray at Lacasdale, in July 1979, that, in a layby since erased by EU-funded blacktop, I proposed to Helen, tying round her finger a handy length of fly-dressing wool (on St Kilda, where there was only ever one metal ring, betrothal involved a strip of Soay sheep wool). Love in a Time of Fishing. Strangely, her professed interest in the pastime that day seemed to wane almost immediately thereafter.

As you head down the other coast in the direction of Rodel – where, in the pre-Reformation church of St Clement's, lies buried the Jacobite warrior Donald Macleod, who at the age of seventy-five married his third wife and 'leaped like a salmon' into her bed (they produced nine children) – you see on your right, lapping against a man-made sea wall, a mercurial little loch named Fincastle. Here I once hooked an eel on an Elver fly, and, during a lunch break, was taken out in the boat by the ever-vigilant factor, Tony Scherr, who had detected a slight change in conditions, and caught two salmon to a Pennell while the cashmere-bellied house guests were still munching their beef and horseradish baps in the boat house.

Fly choice here used to be simple – anything black, unless you had something darker. At the top of this system is Laxdale, the smallest salmon loch in the world. It took me thirty years to land my first fish out of it.

There is a Gaelic saying that clots my throat: 'Three things beautiful in death – a black-cock, a sea trout, and a little child.' Some folk have developed such a fetish about salmon numbers that they count the lithe and delicious sea-going strain of brown trout as an also-ran; but to me they are the most elegant of all our fish, and wonderful when cooked off the hook over a bankside fire, with bacon and a little butter swinging in the pan. *Salmo trutta trutta* is made, not born – in some brownies the anadromous urge takes them out to strengthen and change livery in the salt, returning as silver runners that sometimes you can persuade to dance at your fly. Set one beside some southern stockie and you'd be comparing Pegasus with a pit pony.

Herling, sewin, harvest peal, black tails – call them what you will, they are intriguing and still quite mysterious fish. Science has lavished relatively little research upon their behaviour, though certain anglers have made a deep study of them, especially in the Welsh rivers where much of the fishing begins only when the light starts to falter towards dusk. I have not experienced much of this kind of sport – when the night becomes as black as a Bible cover, I am minimally competent, and have admiration for those truly expert in this remarkable branch of the sport.

Perhaps compensating for my dud safari to the Falklands, I made a trip in 2003 to mainland Argentinian Patagonia, in search of *la trucha plateada* – a descendant of the brown trout introduced there a century ago. I was to fish the Río Gallegos, which runs below a morainic escarpment through this region of scrubland and pebbly basalt, coarse yellow grass and thorn bush (it's no surprise the many Hebridean emigrants to this land felt fairly at home down in the kingdom of the winds). The sunlight was the colour of beeswax, and the water seemed unfavourably low.

Las Buitreras lodge had access to 19 miles of river, and there was an unrelenting downstream wind, rebuffing my casts with water witches and spume twirling off the surface of some of the pools. Yet in places there were many fish – black-speckled chrome beauties that at times charged your fly, and at others gave a deceptive nip, though all surged away with the ocean strength still upon them (*poder del mar*). I was lucky enough in my draw of zones, and finished up the week with seventeen fish over 6 pounds, including one of 20 (taken on a Prince nymph) and another of 22 pounds, that whacked a waking dry fly, and remains a fish of my lifetime. Later that year, I was being rowed up a Hebridean loch by my friend Jimmy, and I tried to tease him: 'My heaviest sea trout this year weighed 22 pounds. Do you think today we might do even better than that?' Without breaking the rhythm of his oar strokes, and resolutely unimpressed, he replied, 'With fishing, all things are possible.' And there you have it.

*

When the night wind ladles rain against the glass and a grey storm flails to get in, slogging the chimney and yelling like a fishwife's curse, the croft house is snug with peat smoke and I like to sit in the window nook and set up my fly-tying equipment, rolling my own for the morrow. From time to time, when there was a vacancy at short notice on one of the castle lochs, we would take a call from Kenny Morrison, the head keeper, which made for tremendous anticipation, and an early start (beginning with that 2-mile walk to the cars).

The spellcheck-challenging Amhuinnsuidhe Castle lies on the road to Scarp, an islet that made history in 1934 when a resident gave birth to twins in two separate counties, in different weeks, and that same year Gerhard Zucker (later responsible for the V-bombs of the Blitz) wept openly as his experimental mail-delivery rocket exploded, scattering thirty thousand envelopes into the sky. The castellated baronial edifice was constructed in the 1860s with sandstone shipped up from Ayrshire for the 7th Earl of Dunmore to impress his fiancée, daughter of the Earl of Leicester; she broke off the engagement, saying her father possessed henhouses that were larger, and the project bankrupted her intended. The property passed to the Scott family. The spectre of Lady Sophie Scott, a former chatelaine, is said quite often to revisit her old bedroom, but on the two occasions I stayed up late there with a flask of malt she did not put in an appearance.

There are three river systems and seven principal lochs to fish. The castle itself is situated on Leosavay ('the bay of light') and from one turreted bathroom there you can see salmon leaping like Jacobite warriors as they wait to run up the Eaval cascade pools and into Ladies Loch. While staying in early September I once took two grilse on a mini-Sunray before breakfast, but the Doctor claimed they did not count, as he was down south at the time. Generally, we have fished it together since 1979, and I must record that, after the initial start-up inertia of our relationship, while I was patently not worthy of joining his clan, we have proved near-ideal boat partners, since he is a left-hander, shares my penchant for Caledonian natural history and the peatier malts, and instinctively knows when one should keep silence. Over the years, our store of mutual reminiscences has accumulated like a cairn. Being a Highlander and a medical man, he is a great favourite with the islanders. Alec, a veteran castle gillie, was once told by my friend Graham Swift that he was going to have lunch with the Doctor and me: 'Oh, he's a fine fisher,' responded Alec, adding, after a tantalising caesura, 'Doctor Fraser.' I'm arguably still not good enough for his daughter, four decades later.

We have enjoyed some notable days on the castle waters, often in the company of Kenny, a stocky, genial man with a roguish smile, who was so strong he was said to be the last stalker who could single-handedly 'hump' out a gralloched red stag carcass off the hill (those island beasts can weigh 150 pounds). We wet the head of many fish with the help of the trusty Admiral, although in retirement Kenny unexpectedly took the pledge; the last time we visited him, I unwittingly brought him a bottle, which he stowed beneath the stairs, 'just in case', before his wife Effie could see it. We traded memories, the old fire came into his eyes, and he said, 'Doctor, those were great days.'

Probably my favourite loch is Ulladale, which entails a longish, spectacular hike until you arrive at the amphitheatre buttress of Sron Ulladale, which overlooks your day's activities. In July 1982, we had a fine session there, the surface knapped by a hard breeze, the salmon lying with their chins against the far boulders of the lee shore, Alec heaving at the oars. I was using a Donegal Blue on the dropper (more than two flies in a good wind disconcerts me), and I pulled out two fish in consecutive casts from the same gap between the weeds. When a third came at me, I struck wildly and tore it loose: 'You broke its jaw!' howled Alec, normally a soft-spoken man. We killed eight salmon to the boat, lugging them all back uphill in a modified plumber's tool bag. On another occasion, the Doctor was trying to impress Kenny with his natural knowledge, and pointed out a white deer on a distant summit. Kenny coughed into his cigarette. 'That, Doctor, is a sheep.'

Most of the lochs will hold stocks of sea trout and salmon at the same time, but I feel it is a mistake to fish mechanically for both species at once. I prefer a longer, slowed-down retrieve for salmon, though you may have to speed this up if you are drifting very rapidly over the holding ground. That exotic Irish pattern, the Goat's Toe, used to be popular here, but these days I'd prefer a Kate McLaren Muddler. When targeting sea trout, an Intermediate line can be handy, biting slightly into the wave like an old silk version, and also getting down to them when becalmed. We have made some good baskets of sea trout from Scourst, a deep and dour loch where I like to dap – although my father-in-law has a tendency to cast his Dark Mackerel at anything which swirls to my fly as it stumbles and pirouettes downwind of us. In 1987, Scourst produced the record salmon from these lochs – a 19-pounder, taken by Dounreay nuclear physicist Donald Carmichael, who later said, 'Tonight I have a greater feeling of achievement than I ever got from a lifetime of scientific research.' Its template hangs on the tackle-room wall.

Just one mile square, Voshimid is, for its size, the most productive salmon loch you could hope to find. Its shoreline is wildly indented, splashes of mulberry-coloured bell heather and blood-drop reflections of mountain ash relieving its stern surroundings. When there is a good grey wind out of the west, with no cat's-paws, and fish in shoals have settled on a falling spate, this loch can be generous, and is as near to piscatorial paradise as I have yet discovered – I was even once rowed there by a gillie named Adam. The poet Louis MacNeice, who wrote a dyspeptic account of his Hebridean travels (*I Crossed the Minch*, 1938), seemed to like nothing much about the islands apart from a blancmange served at the Rodel Hotel, but even he, walking near this loch, had to concede it was 'A blessed place ... at the end of everything.'

It was while fishing Voshimid with stalker George Ross that J. M. Barrie conceived the idea for his whimsical drama *Mary Rose*, which concerns the ghostly double disappearance of a girl who steps ashore on one of its islets (I have never dared do so). Barrie, whose stepson, the inspiration for Peter Pan, was killed in the Great War, enjoyed considerable success with this play when it was first produced in 1920, as it resonated with those who longed for the reappearance of the lost generation, and it is still affecting when occasionally restaged. Alfred Hitchcock scouted for locations to film it in 1964, but the project was never realised. The Hebrides have inspired a peculiarly broad spectrum of writers – from Gaelic poets such as Sorley MacLean and Iain Crichton Smith to Compton Mackenzie, a celebrated nesophile who lived on Barra and once dashed off a novel of a hundred thousand words in thirty-one days flat. He said he consumed half a ton of tobacco during his literary career, and my favourite of his thumbnail sketches is a description of Adolf Hitler as 'that non-smoking lackey of death in his lavatory-attendant's uniform'.

Some days Voshimid is bristling with fins around the serrated lies of Murdo's Teeth, or the mouth of Sopwith's Bay, yet nothing will take – Alec's unimpeachable observation to me was, 'It has passed over the understanding of many men, what makes fish behave as they do.' But sometimes they do co-operate, as we found one afternoon when there was a steady lop on the water along with an intermittent smirr of rain, and salmon seemed to be lunging at our flies on every drift, congregating around the outflow to the little river and along the shallows of the green bank, arcing at the dibbled dropper right up to the oar's tip, with Kenny repeatedly rowing us into the middle of the loch to avoid disrupting the honey hole whenever one was hooked. The Doctor and I took away five fresh fish each, dropping them off at the Harris Hotel for the next morning's bus to take up to the Stornoway smokehouse (several of them mysteriously disappeared en route). That will never happen again, because now, quite sensibly, the majority of fish are returned.

Later, when I stepped outside Molban with my nightcap, the clouds had settled on the sunset like smoored peats, and I thought how passionately I wanted to continue to be part of this place, with its torn horizons, until time is no longer with me.

In 1991, Ted Hughes was staying at the castle with a party from the West Country ('they're not really friends; they are people I fish with'). When we arrived for the day, he was ambling back from the fly-only sea pool toting a spinning rod and a biggish Toby lure. He was a tremendous taker of fish by many methods – later that day he was trying a tarpon streamer on the lochs, and when he saw my new, golden-finished Hardy Sovereign he crowed, 'That's just a tart's reel!' So greatly was his passion for the world suffused with fish and fishing that I feel it is hard to appreciate his work without taking this into account.

I was first introduced to the Poet Laureate by that urbane dry-fly maestro Dermot Wilson, and we fished the Wiltshire Avon together (the first time I caught a glimpse of Ted's unusual fly box, the patterns hairwinged with pubic hair donated by various female admirers). Although I was never one of his intimate coterie he was always a generous spirit to me, and even contributed a foreword to one of my books. Later, we shared some of the places he loved in Devon, and he came as my guest to the annual Flyfishers' Club dinner, where he considerably raised my meagre literary stock by arriving late at our Savoy Ballroom table for all to see. We began at once to discuss a recent book about menstruation (I'm not sure Ted had a concept of small talk), and not only was he a fabulous raconteur, he was also a fine listener. You had to be on your mettle, though, as in conversation he took few prisoners. Eager to sound fascinating and lyrical, I was invariably either tongue-tied or facile, and ended up asking him questions about the weather instead. The celebrated animal magnetism he exuded was real enough, and for me it was like sitting next to Ovid, or Yeats, so great an admirer was I of his verse that struck fire from the nation's flint in an era of the greying of language. Like Norman Maclean, he was 'haunted by waters'. It is an animating principle of his *oeuvre*, welling up in potent hydraulic, hydrolatrous images, and irrigating his prose and poetry, accounting for much of its mystery and depth. The fishy dimension was integral to his apprehension of the natural world, with its quickening components of shamanism, superstition, fertility lore and 'violence' that was more likely to feature the god Pan than gentle old Izaak. I would say he was possessed by fishing.

On the Sabbath, when Scotland forbids angling for migratory species even by holders of the Queen's Gold Medal for Poetry, he and Graham Swift came over for family lunch at the croft. Some days the elements are in a circus, the wind stravaiging round the headland with birds in its hair, but that day there was an overcast calm and the afternoon sea breathed luxuriously. Responding at once to the rough magic of the place, as he inspected the rock pools and frowned out approvingly across the Minch, Ted said he could see why I liked writing there: 'It's like Prospero's cove.' He repeated that phrase in his letter of thanks to the Doctor, referring to it as a 'paradise'. Neither would have been casual allusions. He had just completed his visionary, if in places opaque, study *Shakespeare and the Goddess of Complete Being*, in which *The Tempest* features prominently (a play that addresses notions of Arcadia), as it did in his relationship with Sylvia Plath, author of *Ariel*, who had originally thought of entitling her first volume *Full Fathom Five*. The image of a spell-binding magician conjuring wonders from the water is appropriate to Ted himself, in several enchanting ways, and might pass as a metaphor for the act of angling, while we're at it.

It was in the rod room of the Lochmaddy Hotel, on the neighbouring island of North Uist, that I first heard a sportsman explain that he was in search of Paradise. Lower-lying than Harris, the interior of Uist is more water than terra firma, and fishing on the lochs and sea pools was allocated according to a strict roster system, so we learned to hot-foot it from the ferry to get our names high up on the list. The gillies were all freelance, and included at least one ex-Gorbals gangster wearing a cravat and a poured concrete grin, though the first time we stayed we were hosted by the son of the laird,

Fergus, a tousle-haired, power-smoking enthusiast who had press-ganged his father's elderly retainer, Hugh, to row for us. Unusually for a Hebridean, Hugh wore a tweed coat (most of that lustrous, aromatic homespun is exported); he was an elder of the church, a staunch advocate of the Blue Zulu fly, and a naturalist of the old school, who gravely assured me that cuckoos hibernate underwater.

Some find Uist remote and dispiriting. Bonnie Prince Charlie had a melancholy experience when on the run there in 1746, disguised as Betty Burke and clothed in filthy clouts, living off drammach (oatmeal and sea water washed down with brandy). Although he was a wastrel, Celtic culture identifies with the lot of the exile, and he is still fondly remembered. Later, the explorer Robert Buchanan described it as 'A lonely outer region not dear to the gods ... [where] the wild-goose screams overhead, and the ice-duck haunts the gloaming with its terribly human "Calloo! calloo!"' To the loch angler, though, such overcast conditions would be ideal. Indeed, I have so greatly savoured the wilderness cure afforded by the Uist lochs that this was where I went to celebrate my fortieth birthday (much to the mystification of my Lady of the Layby), persuading the louche young laird of Griminish to row me up to Burn Bay at the head of long Geirann Mill on an October morning of wet and breezy greydom, and there, with the stags belling through the mist on the umber moorland we rose two rusty cock salmon to my Golden Bumble. We finished up by bagging a brace of snipe shick-shacking their way across the pale bog, their breast

feathers frosted with brine from having skimmed the wave-tops migrating under the recent full moon.

One almost unique feature of the fishing here is the series of estuarial sea pools where, with the right guidance, you can access sea trout as they nose up along the tide in search of freshwater outlets. This is true fish-hunting, but it's rare you strike it right. The period just before and after low tide, when the fish are briefly concentrated in well-defined pools, can prove ideal; I like to locate the shoals with a smallish Flying Condom spinning lure, but the purist Doctor decided those don't count, either.

We have had one bonanza at Ardheisker (keeping five nice sea trout), but it is Vallay Sea Pool that I find most attractive. A Victorian linen magnate from Fife, the pioneering archaeologist Erskine Beveridge, built a fine mansion on Vallay island, but his son had an ill-starred liaison with some local lassie and was found drowned in the sea pool that runs through the sandflats below the house. Now just its shuck remains: the home of the archaeologist has itself become a relic, the haunt only of rock doves, but a peculiar air still pervades that shoreline – you steal your fish from the fleeting tidal stretch and give thanks you are free to leave. I know of few other places I fish (Iceland, perhaps) that feel so often numinous as these islands do.

When the BBC was making a series about *Country Life* magazine and some of its contributors, I suggested North Uist as the location for our three-day fishing shoot, feeling certain the deep allure of these furthest reaches of the kingdom would prove photogenic. The sun duly cast its silver net over the dark folds of the sea, the landscape acted its socks off in the scalding glare from skies as blue as harebells, and a glorious, riviera calm prevailed – hopeless for angling, of course, and none of my learned pieces to camera about the evolution of the Muddler Minnow or the history of tweed weaving compensated for the total lack of 'rod-bending action' (they never made the final cut, anyway). For the climactic cooking sequence I was reduced to baking a minute, foil-wrapped brownie. The crew were as unimpressed as poor Dunmore's fiancée. The day after they left, naturally, we made our biggest basket of sea trout for ten years. So it goes.

In the lemon-grey aftermath of a long storm, I climbed up to the House Loch after children's bath time, with my 4-weight 10-footer. The evening air was tinged with salt, and honey from the sparse August heather (you seldom get those wine-dark hillsides of the mainland), and a couple of herring gulls were skulking on the inland fresh water – a sure sign of more tempestuous weather to come. I flung out a hairy great Soldier Palmer, and was taken by a dark, heavy trout that bored doggedly and proved to weigh 2¼ pounds – my best ever from this strange water. I still feel it might harbour larger specimens, though this was a good while ago now.

I was taking tea and scones (the traditional *stroupach*) with Rachel Macleod one day, and she told me this: 'When I was a girl, there were more than one hundred people living here on Cluer Point. We would walk to school barefoot. And within my father's memory there were six brothers living around the loch in their separate houses – we know it as Lochan Caorunn, the little loch of the rowans – and one evening, after feeding his family a meal of shellfish, Pannaigh saw from his door the water horse come up out of the loch and start to rummage around, you know, in the heap of empty shells. And so the man went in, and he said, "That's it for us now in this place." And that whole family – they left on the next boat, to Manitoba.'

The water horse or kelpie is no joke in island culture – a malevolent supernatural being that drags humans down underwater and tears out their viscera. Since hearing that tale, I have twice seen towards dusk immense bow-waves that could never have been made even by otters, let alone trout, and on one of those occasions I actually refrained from casting into the fading light. With fishing, all things are possible.

Calloo.

9

OF MONSTERS

'There is no deformity but in monstrosity; wherein,
notwithstanding, there is a kind of beauty ...
For Nature is the Art of God.'

Sir Thomas Browne, *Religio Medici* (1643)

My mother had been still in her teens when she married Baron
Frankenstein, which is partly why I had an early familiarity
with monsters. She had already appeared in a dozen other
films before she played Elizabeth, the misguided scientist's
fiancée, in James Whale's genuine classic *Bride of Frankenstein*
(1935), and later that same year she starred as Lisa Glendon,
wife of the lycanthrope Wilfred, in *Werewolf of London*. From
the age of six, I also became acquainted with the teratolo-
gical world through the twice-daily sessions of Bible study

I enjoyed with my governess (a Christian Scientist aptly nicknamed Laudie), and this introduced me to Leviathan, Behemoth and the creatures of the Apocalypse, imbuing me with an abiding penchant for grotesquerie which, coupled with the usual exposure to fairy tales and the darker *Märchen* tradition, has since convinced me that even if not every footbridge has a troll lurking beneath it you should be careful how you plan your routes in life, especially around rivers.

Most children have an edgy fascination with monsters, and it is surely no coincidence that we anglers, being nicely childish, cleave to mystery, and the essential notion of prodigies, including of course the prospect of encountering something gigantic with fins. Monsters seem to rear from the sea floor of our collective imagination, repressed, unsettling, yet curiously desired. They form the staple of myriad heroic contests. In Latin, *monstrum* includes a glimmer of the divine, and certainly does not have to connote ugliness. Nor is hugeness a prerequisite: the spider family, and its cousins the tick tribe, are as much evidence to me of God the Creator's occasional perversity as the fabled Burach-Badhi of Perthshire (where I live), a nine-eyed eel that sucks blood from your legs at a local ford.

Hollywood, the internet and the gaming industry have refashioned our visual guidelines of the monstrous with numerous creature features and gargoyle-like graphics, but it's worth remembering that in Mary Shelley's original 1818 novel, the world's most notorious creature bore little resemblance to the bolted, flatpack appearance of Boris Karloff (a gentle giant in real life), being instead athletic, with lustrous, flowing black hair, pearly dentition, plus linguistic fluency in French and German. Nameless, he tells Victor he should be known as Adam, and he learns to read from a purloined copy of *Paradise Lost*. Though animated by celestial lightning (it may be his creator's name partially derives from Benjamin Franklin, then recently famous for his electrical experiments), his very existence challenges the classifiable harmony of Eden, and it is a moot point whether or not Dr Frankenstein himself qualifies as the real monster. Our ambivalence towards monsters and their relation to what we perceive as the natural world is a theme running from *The Tempest* to the *Rocky Horror Show*.

Outlandish creatures have been a continuum in human narrative since classical times – minotaur, basilisk, hydra, manticore – and via dragonish legends and various Bigfoot myths we reach the present-day passion for cryptozoology. Pliny the Unreliable swore he had seen an Egyptian centaur preserved in honey. Illustrated mediaeval bestiaries featured such bizarreries as the female Chichevache, who was perpetually thin since she fed only on patient wives. The phenomenon is practically global: Japan has its Kama Itachi (an invisible, sickle-wielding weasel) and the Malaysian Bonachus produces bowel movements that cover several acres. From the Renaissance period onwards, the *lusus naturae* (or freak of nature) became a staple of travellers' tales, and what is remarkable is how many of these reported prodigies were aquatic (Caliban himself is described as 'fishlike'), and how the belief in water monsters is central to many cultures. The pedigree of Leviathan – who 'maketh the sea like a pot of oint-ment' when he stirs – includes Lotan, the Phoenician dragon, and a Nilotic crocodile deity. Caesar Augustus decorated his villa with sea-monster bones (presumably dinosaur frag-ments). In *The Log from the Sea of Cortez,* 1951, John Steinbeck compared an ocean without nameless monsters with an entirely dreamless sleep. There's frankly no escaping them.

When it comes to maritime monsters, nothing much sur-faces in the collective Western imagination more enduringly than 'a white-headed whale, with three punctures in his starboard fluke' – Moby Dick, referred to by the *Pequod*'s crew as a Leviathan and monster, vengeful, apparently immortal and possessing freakish powers. One of the subtler aspects of Melville's novel is the way the hunted cetaceans are also perceived as marvellous and sometimes pitiable (those incar-nadine descriptions of spouting dark blood), and again there is some question about which is the more truly freakish – the deranged Captain Ahab (and his English counterpart, Captain Boomer, whose whalebone-ivory arm terminates in a mallet) or the albino giant that is the object of his vendetta.

Anglers have a keen appetite for the marvellous, and the literature is well stocked with fish turning inside out, falling from the skies, copulating with goats, creating the universe, hatching from beetles, bursting into song, turning into women, being converted to Catholicism and other unlikely feats. Walton has been chided as credulous for including such wonders as the 'whirlpool' fish and the adulterous Sargus, but he was only reflecting a contemporary taste for novelty and prodigy, as enshrined in the collections of the rich and gullible, who paid sometimes astonishing sums for rarities like unicorn horns which, fashioned into drinking goblets, could allegedly neutralise poisons. (The unfortunate male narwhal, from which such trophies were torn, uses his tusk to test the salinity of water, attract mates and stun char.) Even the learned Sir John Soane in 1698 displayed in his collection a Vegetable Lamb of Tartary – the provenance was sometimes Borametz – which supposedly grew from a gourd. In Antwerp, there was a thriving cottage industry constructing Jenny Hanivers – desiccated skate or ray parts contorted to look diabolical (the nostrils converted to eye sockets) and sometimes customised with African perch tails, to resemble mummified marine monsters. The great taxonomist Linnaeus was hounded out of a Lowlands village for doubting the authenticity of one such confection. You can still buy similar souvenir examples along the Gulf of Mexico.

An associated trade (the hub for a while being nineteenth-century Japan) was responsible for the production of reasonably sophisticated counterfeit mermaids, chiefly hybrids comprised of monkey-cum-cyprinid components. The most famous example came to light in 1817, and was widely exhibited as the Feejee mermaid by the showman Barnum, until lost in the fire that destroyed his museum in 1865. Touring 'raree shows' of freaks being once fashionable, the putative corpses of were-fish and merfolk were much in demand, and were still believed genuine well into Victorian times. One Regency vicar, Robert Hawker of Morwenstow, arrestingly masqueraded as a mermaid on the rocks around Bude for several days, until, still wearing his oilskins and long wig, he was coaxed away by concerned parishioners.

Fish-tailed human figures feature in iconography from the Esquimaux to the indigenous peoples of South America, and include Noah and his wife, the Philistine Dagon, the Blue Men of the Minch who play shinty by moonlight, an early avatar of Vishnu, and Kuahupau, a Hawaiian shark goddess; but the mermaid herself (like the siren, with whom she was sometimes conflated) belongs to a longish line of femmes fatales resulting from the lustful misogyny of seafarers so salt-struck they might yearn for the dugs of a dugong, and who wistfully dreamed of a lewd race that was perpetually lubricious from the sea, but elusive enough to be safely a source of fantasy. Occasionally, some were caught in nets. A celebrated, and peculiarly specific, account from Borneo in the eighteenth century records the mermaid Amboina being kept alive for 'four days and seven hours', during which she squeaked like a mouse, though her 'excrement was like that of a cat'.

Not all *femmes poissons* conformed to the voluptuous stereotype – some were reported as simian and bristle-headed – but on the whole the associations were lascivious, typical of which was that briny folk ditty 'Paddy Miles and the Mermaid' ('and what should have been mutton was nothing but fish'). Until the late nineteenth century, Spanish sailors had to swear in front of a magistrate before embarking on a sea voyage that they would refrain from intercourse with mermaids. Given the technical impossibility of intercrural sex, this would presumably have covered the prospect of alternative forms of gratification: oral, aural, axillary or by manustupration – a dilemma neatly solved by Magritte in his 1935 oil painting *Collective Invention*, which depicts a mermaid with inverted particulars, appending a gadoid head above human nether parts. The pervasive identification of fishiness with sexual aspects is chronic and ancient: 'fishpond' was mediaeval slang for the vagina (which explains those adulterous adventures of Little Tom Tittlemouse, for example), and in modern gay parlance a 'fish' signifies a heterosexual female, as well as a homosexual sailor cruising the bars. But, I digress.

These fishy female figures once created a moral quandary for the Church (though perhaps not in the Bude area) as they were occasionally, in maritime chapels, for instance, represented in ecclesiastical murals alongside St Christopher, or on carved roof bosses, sometimes provocatively bifurcated. Their pagan iconography may have been uneasily incorporated into Christian myth, but since they were heathen creatures, could they be saved? One Irish legend tells how in 558 a mermaid named Niban – a maiden who had been swept away by an overflowing sacred well some three centuries previously – appealed to St Comgall and was netted; she converted to Christianity, and at her death ascended to become one of the host of heavenly virgins. More commonly, though, oral tradition explains how merfolk (the male of the species was considered more violent, and less fun) are fallen angels, and therefore to be avoided as uncanny, intent solely upon temptations towards sin. This did not deter the crew of the good ship *Halifax* when, off Mauritius in 1739, they allegedly captured several merfolk which averaged two hundredweight each, the females weeping as they breastfed. Apparently, they tasted agreeably like veal.

After Frankenstein's creature, our most famous monster is Nessie, first chronicled back in AD 565 when St Columba did battle with her (I always conceive Nessie to be female). General Wade's troops constructing the Dores–Foyers road nearby in 1726 several times disturbed mysterious 'Leviathans'. Other lochs in the Highlands have less renowned but equally entrenched monster histories: Morar (at 1,080 feet the deepest in Scotland – far deeper than the surrounding sea) is home to the similarly saurian A'Mhorag, and Shiel harbours the resident Seilag. They all seem to put in regular appearances when times of peace are about to be fractured by violence.

In the 1930s the Doctor's great-aunt Netta bought a boarding establishment (Half Way House) on the north shore of Loch Ness, and before long she and her sister happened to spot the monster, which certainly brisked up business. She (Nessie) had previously been sighted in 1932 by Colonel Fordyce, who described a camel-like shape, and from the tea room at Half Way in 1933 she was also espied by the Revd W. F. Hobbes. In 1934 Arthur Grant claimed to have collided with a 'plesiosaur' on his motorcycle while it was making off with a sheep (this was later reckoned to have been a dog otter bearing a fish, as Mr Grant confessed to having the drink taken at the time).

Urquhart Bay remains a favoured spot for sightings, and here the later discredited 'surgeon's photograph' was taken. Whether eel, catfish, freak wave form or pinniped (there's an extensive mythology about selkies the further north you go), Nessie is the *ne plus ultra* of the denizens of our deeps.

Gaelic tradition is underwoven with bogles, sprites and shape-shifting *bocan*, none of them to the slightest degree figures of fun. The Harris water horse was the *each-uisge*, a still-water spirit animal occasionally adopting human guise, with a penchant for seducing local maidens (one telltale sign would be sand in the handsome youth's hair, though in a Hebridean community this might apply to practically any active male). They were commonly thought to inhabit Lochs Oich, Lochy, Arkaig and Assynt, among others, and in Shetland they were known as *shoopiltee*. I suspect these are vestiges of some earlier Celtic horse cult. Certain Victorian sportsmen even equipped themselves to hunt one down as a trophy for their gun-room wall. Its counterpart river horse was the kelpie, more murderous by reputation. A smith from the isle of Raasay, discovering just his daughter's lungs left on the riverbank, lured the kelpie from its watery lair with a roasted sheep, grappled its shaggy coat with specially forged hooks, and slew it; all that remained of its carcass was that jelly-like substance called 'star shine'. (The aetiology behind some of these myths is remarkably intricate.)

Most aquatic monsters of legend had some deformity that prevented them from seeming quite beautiful. The *direach* of Loch Etive had a hand growing out of its chest, another lacked nostrils. There was a Celtic mermaid, *ceasg*, that sported the tail of a grilse (another good one for the lodge record book), and the fearsome *luideag* hag from Skye was rag-like and dragged the unwary beneath the wavelets of the Lochan of the Black Trout. In the kingdom of the Gael, you were well advised to be careful what you fished for.

The *sith* were faery folk who formed the secret commonwealth, and were frequently associated with salmon and trout. In his numinous poem 'The Song of Wandering Aengus', Yeats (a lifelong angler) imagines an enchanted trout transforming into an elusive 'glimmering girl', who beguiles the narrator into pursuing her for the rest of time. The *sith* were used to explain a spectrum of apparently unnatural occurrences, from infant mortality (and behavioural misdevelopment – a 'changeling' substitution) to epilepsy, sudden rheumatic attacks, or the bruises resulting from wife-beating. Being afflicted with paralysis was commonly attributed to being elf-shot by a faery arrow (from which idea we derive the effects of a 'stroke'). Comparable 'hidden' creatures recur in Scandinavian culture, too: Iceland, birthplace of the Kraken, has the marine *Martroll*, the *Huldufolk* (invisible, unwashed children of Eve) and the water horse Nykr (from which Old Nick himself gets his nickname). In Bíldudalur there is even a Sea Monster Museum. These parallel, powerful populations formed part of the Celtic and other ancient systems of belief long after the coming of Christianity. I have heard it suggested that the sightings of such other-worldly creatures were so common because many Highlanders lived in a state of perpetual malnourishment, and were therefore liable to hallucinations, but, despite my rigorous early Bible training, I'm more inclined towards the animistic belief that there may be several kinds of person in this world, not all of whom are human. (As Palestinian storytellers used to conclude: 'This is my tale, and in your hands I leave it.')

Anglers are naturally drawn to jumboism, and the concept of the outsized is integral to many of our days spent musing over the water. Good things sometimes do indeed come in large packages and, while I am wary of invoking that hoary old 'hunter-gatherer' analogy (too often a convenient excuse to behave unthinkingly), it's true that in those early days our ancestors knew that attracting the favourable attention of Mrs Caveman was more likely if you came back lugging a suppertime sturgeon as opposed to a mere skinful of smelts.

The vernacular lexicon to describe those Brobdingnagian trophies (how Swift himself would have appreciated the impulse) is now elaborate – Fishzilla, Moby Pike, stonker, lunker, hawg, Goliathan, Harvey Wallhanger – but I don't care for those that demean our quarry (brute, or lump). Overdue a revival is the colourful 'sockdolager' – originally a pugilistic term for a sudden knock-out blow, but coming to signify an exceptional fish (Twain uses it for Huck Finn, and so, deftly, did Patrick Chalmers). Another favourite of mine is the Scandinavian *monsterlax*, for a truly gigantic Atlantic. I have never yet landed one, but I know several chaps who have grassed salmon in excess of 50 pounds; yet, even in Fred

Buller's compendious volumes *The Domesday Book of Giant Salmon* (2007/10) no other account quite seems to improve on the romantic saga of Miss Ballantine's Record Fish.

The sun hangs low over Birnam Hill on the evening of 7 October 1922. Mr James Ballantine, professional 'fisherman' for the Lyle family (of Golden Syrup fame) on their Glendelvine beat of the Tay, is rowing his daughter Ina in the Boat Pool, just above the Bargie Stone (Melvin, his assistant boatman, having knocked off promptly at five). Around six-fifteen, the artificial dace bait she is trailing from her greenheart rod is taken with a violent 'rug', and a powerful fish races 500 yards downstream towards the old Caputh road bridge, 'leaving only a whirl of spray in its train' as they hasten after it. It never once shows. 'Dinna let the beast flee doon the watter like that,' admonishes her father, perhaps unnecessarily. He refuses to help her with the rod. Eventually, she manages to steer this fish below the bridge, where it sulks deep in the gloaming.

From Victoria Cottage, her mother comes down anxiously with a lantern. Ina demands a new dress if she succeeds in subduing this monster. 'Get ye the fish landed first,' says James, 'and syne we'll see aboot the frock.' The angler seems to be tiring more quickly than the salmon, and she suggests they budge it from the lie with some lobbed stones. 'Na, na, we'll try nane of thae capers,' replies her father. Several feet down in the dark, her adversary continues jagging. It is now twenty past eight. The paternal 'fisherman' knows the gut trace he has constructed measures precisely 3¾ yards in length, so with his gaff he gingerly feels down past each of the blood knots and judges how deep the 'refractory beast' is lying, then lunges expertly and, in one sweep, cleeks the fish and heaves it into the boat – a heroic feat for a man of nearly seventy, as this salmon weighs half a hundredweight. The cock fish measured 58 inches long, and was confirmed at over 64 pounds on the steelyard at nearby Boatlands Farm. It remains the British record, and is now unlikely ever to be bested.

A plaster cast was made by P. D. Malloch, and, after happy days spent on the Glendelvine pools, I have several times gone up to pay homage to the original Beast, magnificent in pride of place on the billiard-room wall. The flesh itself was donated to the Perth hospital, after first being displayed in the window of Malloch's tackle shop. Its modest captor, watching from the fringes of an admiring crowd, overheard one elderly male pundit pronounce, 'Nae woman ever took a fish like that oot of the water, man ... that's a lee anyway.' Miss Ballantine got her new frock, and lived out her days, unmarried, in Victoria Cottage overlooking the scene of her triumph. She was buried in the Caputh churchyard in 1970. I sometimes wonder how Melvin felt, having become forever a size 10 footnote to angling history.

Set against the epic scale of such an achievement, the pursuit of huge, artificially reared stock fish is a modern tendency that can resemble a canned safari. The history of stew-pond breeding for the table is long and distinguished, of course, from the Roman *piscinae* through mediaeval monastic husbandry (the word 'stew' comes from the bawdy houses of

Southwark, originally steam-bath houses, in close proximity to which were located certain episcopal fish ponds). Many of our modern waters would be devoid of life were it not for the supplementary introduction of stocked fish, but in recent years the selective breeding industry has begun to produce unfeasibly large specimens – carp and rainbow trout, in particular – that are out of all proportion to what a smallish water might naturally sustain, and a sometimes spurious machismo attaches to 'stalking' these Bunter fish, some of which are slovenly-looking slobs paraded as fish porn on internet sites, though most print periodicals are now much choosier in the images they publish.

There is nothing wrong with these Frankenfish as a feat of pisciculture, comparable with the admirable cultivation of prize marrows (one overstocked trout lake even used to lend you a wheelbarrow for conveying your catch to the hut scales), and each angler really must draw their own line at what constitutes an individual sporting challenge. In small stillwaters, some stockies are easy to hoodwink, but if they survive the initial period of vulnerability they soon become hard to catch. I used to do a lot of this type of trouting in the 1980s, when the vogue for 'supertrout' began at Avington, Sam Holland's keenly promoted fishery in Hampshire, and once I did land two 'doubles' in one day – rainbows weighing 18 and 13 pounds respectively, great green-backed giantesses with carmine stripes along their flanks. This remains my heaviest brace of trout, but, handsome and fit as they were, and thrilling to target with a little nymph, we should remain mindful that fish of such monstrous proportions are unnatural, and there's no confusing a poodle with a timber wolf.

One version of the Angler's Prayer runs: 'God, give me grace to catch a fish so large that even I,/ When talking of it afterwards may never need to lie.' But sometimes when confronting a vast fish there comes, along with the frisson of excitement, a slight sense of tremulousness, even dread. I hesitate to liken this to an approaching duel, or the prospect of hand-to-hand combat, but let's just say it's somewhere between waiting in a cocktail lounge for your blind date, and sitting in the silence of your dental surgeon's waiting room. I have experienced these mixed feelings when coming in upon a pod of sizeable female tarpon, for instance (those formidable 'moo-moos' seemingly the size of a cow), or a vast, lone barracuda with his five o'clock shadow – an atavistic vacillation that warns you the creature you are confronting is Other – so that it comes almost as a relief when they do not attach themselves to your line. You are unlikely to be affected by this when in pursuit of gudgeon or roach, but certain species seem to possess *terribilità* (many of the creatures in Jeremy Wade's inventive *River Monsters* programmes, for instance). T. H. White described playing a salmon 'lovely and terrible, like a shark', and Scrope fought that salmon which was a *'monstrum horrendum ingens'* of a fish, echoing Virgil's phrase about the raging Cyclops Polyphemus. The much-overworked adjective 'awesome', with its divine nuance, truly is the word to describe certain aquatic encounters. Piscator, beware what it is exactly you pray for.

Along with gigantic eels, pike – *Esox lucius* the 'freshwater shark' – have long been monsterfied, for predictable reasons concerning dentition and longevity. Esox fables include attacks on horses and foxes, barking like a dog, devouring infants and seizing croquet mallets. The 'luce' is historically famous for being able to strike like lightning: when he identified bacteria in 1683, Antonie van Leeuwenhoek described how the microscopic *Selenomonas sputigena* 'shot like a pike through water'. One Irish guide, showing visitors around a castle, pointed out the skull of a lough pike of astonishing proportions. Next to it on the wall hung a cranium rather smaller: 'That,' he explained, 'is the same fish when it was younger.' H. E. Bates saw a pike 'as big as a donkey', and the aptronymically named A. A. Luce had a fishological moment on Lough Sheelin when a 'monster' pike surfaced and 'He looked half as long as the boat' (those boats were typically 21 feet long). But the finest evocation of this visceral ambivalence about a fishy encounter comes from Ted Hughes's much anthologised early poem 'Pike', wherein the lone fisherman after nightfall, his hair frozen to his head, 'dared not cast' at the prospect of what might be rising from the depths to meet his challenge. This is angling as seance – 'Who's there?' If the experience of handling rod and line is quite often strangely more than the sum of its parts (and I believe it is, or it would scarcely be worth examining) then this sort of contradictory mixture of emotions is an example of what marks it out in the realms of invented pleasures.

*

The sharks themselves, once understandably abominated by seafarers, are here in a category of their own. I have never tangled with a great white, but most of us know its modern avatar in the 1975 movie *Jaws*, wherein Sheriff Brody catches his first sight of the fish and says, 'You're gonna need a bigger boat.'

We were in a basic bonefishing skiff just offshore in a Cuban archipelago, chasing some shoals of bonito and albacore – a half-acre of breaking fish-busting baitballs, then sounding, as we pursued them in 'run and gun' fashion – when the universe opened up along one of its seams and a glistening fin the height of a windsurfer's sail rolled up above the sea among the churning scombroids. Christ on a Harley! Some vast animal was gliding slightly below the surface. I had no idea what this was, and Pedro my guide, lambasting me in Spanglish, managed merely to command, 'Dayvee, no carsst!', though I flung my Seaducer at the beast anyhow, because it's not every day you get that close to the world's largest fish, even if it is a relatively innocuous filter-feeding planktivore, with an almond-shaped mouth where no hook would ever find purchase.

The adult female whale shark can grow as long as a London double-decker bus. When later we drifted over that one, marvelling at its pale-starred back, I reckoned it measured around 40 feet. Unlike most of its toothier kissing cousins, it is a gentle giant, and may weigh 20 tons (a bit of a struggle on my 9-weight). The skin is 3 inches thick, but without blubber: as a result, they sink when they die, therefore few carcasses of *Rhincodon typus* have been recovered. No one knows how long they live (reckoned to be about a century, anyway) or why they dive so deep (tagged specimens have been recorded at 6,000 feet). They have never been observed mating, nor giving birth, although one pregnant female captured off Taiwan had more than three hundred pups alive inside her. The immense proximity of the one I saw, making the sea seethe like a pot of ointment, delivered me a sudden jolt of childhood loneliness in the face of the hugeness of the universe.

More people have been in outer space than have ever visited the nethermost reaches of our oceans, and we don't need to look upwards to find aliens. When, in 1960, the bathyscaphe descended 35,000 feet to the 'basement floor' of the Mariana Trench (deeper than Everest is high) and settled on the gritty marine 'snow', through the viewing window were seen previously unimaginable fishes. There must still be so many further abyssal mysteries to be confronted. Therefore, of course I believe in monsters, and besides: 'the facts of the world are not the end of the matter.'

One country where numerous streams and backwaters virtually exude a sense of the wonderful is India. I have fished there several times. On my gap year in 1974 I caught a baby golden mahseer (that 'salmon of the Raj') on a Baby Doll reservoir lure from the Ramganga in Corbett National Park. Since the heyday of the *chota peg* and pith helmet, the ichthyology of the subcontinent has been minutely studied (including in that 1905 monograph by the aforementioned Dr E. Cretin), and identified species include the Malabar ricefish, Indus snowtrout, honeycomb whipray, purple spaghetti eel and spiketail paradise fish. The practice of angling was a broad church, and embraced fishing for crocodiles ('muggers'). In Bengal, one technique was to suspend a live puppy above the water with a large hook between its legs (bad way to go, Kumar), or, for man-eaters on the fabled Cauvery river (once the classic water for mahseer), locals would float-fish a dead pariah dog transfixed with a double-hook rig purpose-forged by the blacksmith. There was even a bankside spear-shrine to the local hook deity, Thundilkaran.

In January 2013, we went to the Kerala region on a family holiday, and I travelled (by agreement) as an unarmed civilian, sans tackle. Wherever we visited, helpful folk would enquire, 'Why sah have no line with him?' I had no float rod for the pearlspot of lovely Lake Vembanad with its masses of water hyacinth, and in the plantation tanks of Coorg no pike traces for the armour-plated snakehead *murral* (once used for target practice by the nobles of Gwalior), although K. K. our host assured me 'last week they netted one that took three chaps to haul ashore'. By the time we reached the Nagarhole National Park, in Karnataka, fin fever had fairly set in.

There was a spectacular drought, and the lake brochured as fronting our camp had disappeared entirely, leaving us just a reduced version of the Kabini river (an affluent of the Cauvery). Anyway, I had no formal *shikari* to guide me. At length, a grizzled bottle-washer from the kitchens was volunteered, and promised to take me and our son James out in his coracle next morning. He equipped us with a relatively inflexible uptide sea rod, not exactly in showroom condition (the broken tip had been bandaged with insulating tape), and the one available camp hook – a size 2 spade eye example, around which was to be moulded a ball of *ragi* (millet flour paste traditionally admixed with turmeric) about the size of an owl's egg. At ten, when the riverside heat already felt like a pizza oven, we three men went astream in a coracle of bamboo and tarred hide for a session of makeshift touch-legering.

Target species had not been discussed, but we anchored in 6 feet of murk just out from a bay where village women went to scour their cooking pots. Amid painted storks and grazing bison, we took turns lowering our offering into the mulligatawny current. On his sandbar, a lone marsh mugger appeared to snooze. Silvery little *chilwa* flipped and frolicked like bleak. There came a few desultory taps at the business end – small school fish, I supposed. 'Nibbells', explained our guide, hunched over his palm-cupped *bidi*. I was still unclear what we were hoping for. 'Ghost fish, sah. Beeg.' Tap, tap – is there anybody there? At the next nibbell I struck magnifi-cently: there was solid resistance, and it was clear I had snagged fast on the bedrock.

'Beeg fees.'

Now, I have fished in enough places to recognise a guide's trick of helpfully imagining fish activity when sport is slow, but as I horsed the disadvantaged rod around in order to liberate our only rig, that angling journalist's ghastly cliché became a reality – the bottom began to move. Sluggishly, with almost insolent progress, the line shifted to the downstream side of our tar boat. The wounded rod bore up staunchly, while I pumped as valiantly as possible, given my crouching posture, and suddenly the hook came back at me, baitless and slightly deformed from its original aduncity. The sensation was similar to hooking a giant skate off Mull, when it clamps its wings onto the sea floor and by suction defies all your leverage. This routine was repeated twice during the next hour, but we could not raise anything far enough from the bottom to make real headway. James is no stranger to huge and strange fish, having worked on a commercial long-liner in the Tasman Sea, but we remained baffled.

'Sah, beeg.' I was just contemplating an infringement of the 1897 Dynamite Act, when it was time to be paddled back for noontide tiffin.

So, what exactly had we been connected to, all unseen and unsuccessful? I spoke on the phone to the resort's co-owner, former international cricketer Saad Bin Jung. 'We have a type of freshwater shark here,' he enthused, 'also snake-headed fish. Some grow up to six feet. You must come again, but bring your own gear.' Although in the Ganges there is a true freshwater Indian shark (its feeding habits benefit from the Hindu custom of water burial) the term here usually refers to the *Wallago attu*, or *gwalli*, a silvery, dagger-shaped catfish with serrated teeth and an impressive array of feelers. It is hard to subdue, and is a favourite among snigglers. Another contender might be the *Silvad*, a more hulking, scale-naked silurid with a blue back, which hugs the riverbed and possesses a carpet-bag mouth. I have caught a few catfish (even once fly-casting for them in the Arno just upstream of the Ponte Vecchio), and once at night hauled out a biggish Zambezi *vundu* using for bait a bar of soap (apparently they are attracted to the palm oil in it), but I was not prepared for the sheer intransigence of these invisible monsters. So, one of these fine days I shall return like some district officer of yesteryear, elaborately equipped with the latest tackle – in this case livebait harnesses, billfish-grade fixed-spool reels, graphene whopper-stoppers and perhaps a recommended supply of small brown frogs – in pursuit of my spectral snake-shark, or maybe a spiketail paradise fish. But we are definitely going to need a bigger coracle.

10

THE CHALK SPIDER

'A man that looks on glasse/ On it may stay his
eye;/ Or if he pleaseth, through it passe,/ And then
the heaven espy.'

George Herbert, 'The Elixir' (1633)

On a lip of the Chiltern chalk ran the nicely derelict, uncouth
stream which I fished every day I was not away at school,
from Easter until the autumn half term. It was unkeepered,
unstocked and so overgrown that a conventional cast was
virtually impossible, and jungle warfare tactics had to be
involved – part of its appeal during my teenage years. Here I
was able to observe a few solitary trout in their element. If the
Itchen wildie is a sleek young cavalry officer, these were the
Bashi-Bazouks of the truttaceous world – roughneck, blotchy

slabsides which inhabited a chalkstream back country that nobody else visited. Jouncing down the half-mile parkland track in my old jalopy, rod sticking out of a hole sawn in the roof, I would have the place to myself. I thought of those fish when I was meant to be doing more crucial things, like playing full back in a second House side, or making notes on the rise of mercantilism. It was here I saw my first kingfisher – that halcyon splinter of the spectrum, without which any chalkstream eulogy would be incomplete – and first caught a big fish on my own. It became Home Water.

In a tangly hole I had christened the Coffin Pool there lived one coppery-flanked warhorse that eluded me. He had a distinctive cicatrice on the back of his skull – probably a heron's spear mark. Over the course of two years I lobbed him the Killer Bug, Polystickle, Docken Grub and natural minnow-tail (my purist phase was rather short-lived). From other lies I had eventually extracted fish up to 3 pounds, but this one never made the mistake. I tried 'bushing' for him with a bluebottle, and even the freelined worm. I mail-ordered a newfangled fibreglass rod kit, and finished the honey-coloured blank with ring whippings and some bright orange varnish (the result resembled a señorita's painted fingernails clutching a panatella, though that analogy only occurred to me years later, in a Cuban nightclub). Then one Boxing Day, out rough shooting, I saw a heron swerve up from the margins, part switchblade part umbrella, and as it flogged away across the morning sky I discovered my great trout's torn corpse on the frosty bank, where Old Nog had finally made a Christmas feast of him.

Below our stretch was a more open beat preserved by the local squire, and he had repeatedly declined permission for me to fish it. There was a low weir, below which covertly I had spotted several nice fish, smug as provincial celebrities. I was too timid to poach it alone, but my father – who had shown little interest in the river before – was familiar with the allure of forbidden fruit (it had famously got him into trouble), and suggested we try our luck together. As we slunk into position, he took off his hat with the jay's feather, and smoothed back his sparse hair, as if in preparation for some public appearance. Indeed, had we been apprehended, it might have given the local cub reporter a useful headline: DISGRACED TORY MINISTER CAUGHT RED-HANDED! But there was nobody else about. Just below the sill, I plipped in my minnow. There was a bronze convulsion, a long pull and the fish tried for the shelter of the willows. But we had tied up strong tackle, and soon we were hastening back up the ride to our own place of safety, above the march fence, my father as gleeful as a joy-rider. It weighed 2, indignant, pounds on the kitchen scales, though the flesh was pallid and poor. We never tried such a sortie again.

By the age of seventeen, I had discovered other distractions – Alice Cooper, Balkan Sobranie, an unattainable Emma – and then we moved away. The river was cleared and re-sculpted; many things changed. But I learned there the essential practice of peep and creep, which has stood me in good stead as an angler these fifty years – the need for stealth – and the great enjoyment of fishing for individual fish you can see, in transpicuous water. (Some anglers actually use rose-tinted spectacles to enhance contrast, though I prefer amber lenses.) With practice, you can improve your ability to spot fish, like a colour-blind sniper unfooled by camouflage. And this experience of looking below the glassy surface, as the poet wrote, can afford you a glimpse of Paradise.

In his decidedly odd short story 'The Curate's Friend' (1911), E. M. Forster described the distribution of our chalklands as resembling the body of an immense 'chalk spider' straddling the British Isles, its legs being the South Downs and Chilterns. With other outcrops from Yorkshire to Argyll, this distinctive terrain – a habitat for beechwoods, orchids, flax – has come to define something about Britain, from the iconic White Cliffs to the pastoralism of our water meadows. Those Downs are really uplands (from the Old English *dun*, a hill), and such cretaceous escarpments also give rise, in their declivities, to dewponds or cloudpools, which is why Jack and Jill went *up* a hill with their pail.

Chalk was formed when much of our present land mass lay beneath the sea, and the armoured remains of innumerable tiny organisms accumulated to create a lithological bed of minerally rich calcium carbonate – a sedimentary monument to marine deaths some sixty million years ago. Groundwater percolates through this soft basement and forms aquifers, which issue as springs when sufficient pressure has built up; they are the cool-water deposit accounts from which chalk streams withdraw their currency. (Temporary surface lavants, or winterbournes, may appear after very wet spells, and such 'landsprings' were once thought to presage disaster.) Unlike our rain-fed moorland streams, these watercourses tend to maintain a steadier height, and carry dissolved nutrients that maintain a lively food web, with dramatic weed life – viridescent starwort, and water crowfoot with white flowers that frost the stream like a tequila glass. Ideally managed, chalk-stream water is indeed as clear and cold as an intoxicating spirit, and about as expensive.

Historically, chalk country has focused a particular image of rural Albion (some 80 per cent of the world's chalk streams are British), often the idyll of a peaceable, almost prelapsarian world that offers an antidote to the havoc of our lives. Charles Kingsley, a Lambourn divine, once preached a sermon on the virtues of chalk streams, and I feel certain Millais' Ophelia is depicted as drowned among tresses of honest English ranunculus. It's a landscape that seems to foster nostalgia. Fishing the chalk has connotations of purity, seclusion and finesse. It may seem tame, but in a world of exotic, rock-and-roll angling destinations there is still much to be said for that gentle plainsong.

As you stride through the meadows, the morning grass straggled with dew, the sun comes stammering through the glade, as pale green as elf light. At the ford, cattle blink, and lick their lips with pumice tongues. A moorhen gives a metallic clack among the reeds. Beneath the far stanchion of the footbridge there is a surface pucker, and you kneel to the river. The line sings out, and there comes the 'gluck' of a rise – Tom

Stoppard compares the sound of a cricket ball struck perfectly by an excellent bat to the noise of a trout taking a fly (*The Real Thing*, 1982) – and you tighten into your first fish of the day, buttercups dusting your fingers with gold as you lift it from the meshes.

But all is not well in Arcadia. The lie of the land is often just that – an illusion. Our chalk streams are not natural – they are the result of careful micro-management over many years – and, despite sterling conservation work, some of them are now in trouble. Abstraction by water companies (chalk water is easy to access) means reduced flows that concentrate pollution, raise temperatures and cause siltation and diatomaceous slime. Some stretches of these once pristine streams are slouching and moribund – dying of thirst. We are borrowing these rivers from the next generation, a perilous debt. I suspect future historians will regard with incredulity how we failed to conserve such a precious commodity as fresh water, proverbially the lifeblood of the planet. We have taken it for granted. Back in Victorian times, Cecil Rhodes once remarked that if cold fresh water cost a guinea a glass, nobody would want to drink anything else.

One of my sporting heroes is Colonel Peter Hawker of Longparish House, who continued to fish the Test on crutches or horseback, holding himself 'as straight as a lance' despite receiving a crippling wound when he was serving as a hussar in 1809 (he fortified himself against chronic pain with Dr Badger's quinine mixture). He was scarcely a purist, but in

1816 he visited Stockbridge – soon to become the *fons et origo* of Victorian fly-fishing strictures – and described the Test there as 'not worth a penny', decrying 'the cockney-like amusement of bobbing with a live mayfly'. In those days, crossline and dapping were such common practices that bankside trees were felled to maximise breeze, and the blow-line was still in use at the fabled Houghton Club as late as 1886, when William Lunn arrived as head keeper. A veritable revolution in attitudes was in the air, however, and from then on the deployment of a natural insect bait anywhere in chalk parishes would have had outraged yeomen beating a path to your door with pitchforks and flambeaux.

In 1872 there waged a bloody feud between the Shawnee and Delaware Indians over the ownership of a child's pet grasshopper, and it always strikes me the so-called war between dry fly devotees and the cult of the nymph has something similarly absurd and unfortunate about it. The idea now seems risible that a set of fastidiously codified protocols involving the exact imitation of an upwing dun cast, floating on the surface, to a rising fish was in some quasi-moral sense a superior form of angling to the dubious merits of fishing a sunk imitation of the insect at an earlier stage of its life cycle, but the arguments were fervently upheld by both sides. In one corner were supporters of F. M. Halford, sometimes unhelpfully typecast as the high priest of floating fly ultra-purism, though he was keen enough on Thames float-fishing as a youth (his boatman Rosewell being a fine example of nominative determinism). I am not much taken with Halford as an author, as his prose can be sclerotic and dogmatic, though he wrote plenty of sense. On the page he seems humourless, albeit anyone who can have had such a devoted friend as G. S. Marryat (a quirky and original *éminence grise* behind much of the era's fly-fishing experiment, though he never sought much credit for his insights) must have been fairly congenial – but perhaps only if you agreed with his precepts, which still appeal to some as the epitome of fly-fishing. It is just a shame that Halford managed to make anything other than his techniques sound caddish, and that a little of this piety has survived into present times.

Ranged against him in this battle of Mice and Frogs (Homer's comic *Batrachomyomachia*) was his former friend G. E. M. Skues, a canny though unusually modest solicitor, who helped pioneer the development of imitative sunk nymph techniques, and effectively became the devil's advocate in the eyes of all those fervent 'dryflydolaters' (the angling historian J. W. Hills says he was excoriated as a 'dangerous heresiarch' in traditional circles). Altogether a more engaging writer, his early volume *Minor Tactics of the Chalk Stream* appeared in 1910, and, according to some accounts, Halford confronted him in London's Flyfishers' Club, exclaiming, 'Young man, you cannot fish the Itchen in the manner you describe in your book' – to which the lawyer is said to have replied, 'But I've done it.' They never again spoke to one another. Skues was not responsible for the deep nymphing that came with weighted patterns devised much later by Frank Sawyer (the 'Netheravon style'), and publicised by his pupil Oliver Kite, who incidentally had been struck by lightning during the war, and he might well have disapproved of some of the plumbaceous modern confections that are passed off as 'nymphs', which would certainly have caused old Halford to be spinning in his grave.

Beelzebub may be overall Lord of the Flies, but according to the Book of Enoch, there is a specific angel – Shakziel – who presides over waterborne insects. Some knowledge of trout chow, bat grub, swallow snacks and spider food is definitely an advantage for the angler, but I must confess to attaining a low level of such expertise. I have friends who are accomplished feather fixers and entomologists, such as the author Neil Patterson, who has made a deep imaginative study of the subject and the design of suitable artificials; he dwells on the Kennet and will set his alarm for three in the morning to ascertain if there is a caenis hatch. To achieve an Immaculate Deception by close facsimile is laudable, but beyond my abilities; I prefer the wartime aircraft recognition guideline GISS (general impression of shape and size) coupled with the mnemonic KISS (keep it simple, stupid), so that in effect you are saying to Madame Trutta, 'Giss a kiss.' For dry patterns I happily resort to the Parachute Adams, Humpy or Stimulator, all of which float like some mediaeval witch. The Daddy is another general standby, once known as the Old Tailor, perhaps for his frantic sewing motions, or his legs resembling a mouthful of pins.

Fly patterns can become a fetish. I like the story of Lord Pouting, who was having a tough time one hot afternoon on his syndicate water and, turning to his attendant river keeper, said, 'Now, Fyke, I think a siesta would be a good idea.' 'Certainly, my Lord,' came the reply, 'or indeed any fly with a touch of red in it.'

It was Marryat who coined the punning phrase, 'It's not so much the fly, it's the driver' (a fly was then a type of horse-drawn carriage), and I believe that presentation is paramount – I only wish I were more adept at it. Casting is like a glass of champagne – your first will prove to be the best. You should use the shortest throw that will serve the purpose, with minimal false casting, and aim for a dragless drift. As that urbane Swiss hotelier, and inventor of the après-ski boot, Charles Ritz stipulated, you match the drift, not the hatch. I feel the dry fly should behave like a ballet dancer up on the surface stage (graceful, precise, as light as possible on its toes), whereas the nymph is a belly dancer, wiggling away between the tables where the trout are feeding, down in the basement. Both, surely, are beguiling and legitimate art forms.

Perhaps the only minion of Shakziel that is familiar to civilians is our socking great mayfly, the largest British ephemerid and a byword for the short lifespan (certain cartoonists make excellent use of it). But in fact it is a double misnomer: not only do some of them live more than a day as adults, the vernacular name derives not from the calendar month (they are often at their height in June, and I have seen them hatching in October) but from the seasonal blossoming of the hawthorn shrub, or mayflower.

There are three types of British mayfly, but *Ephemera danica* is easily the most common, and it is a harpactophage's dream dish. Its appearance occasionally triggers a feeding frenzy among trout, but not to the extent that the hackneyed term 'Duffer's Fortnight' might suggest; I did once, on the Chairman's Beat at Leckford, cut my hook off at the bend after I had caught so many trout, and continued to fish just for the spectacle of the rise, but that was a unique experience. I don't know if Sir Terry Pratchett was an angler, but in his sublime Discworld novel *Reaper Man* (1991) there is a supernatural black trout named The Death of Mayflies, and the insect's ephemerality is a running motif, with the skeletal Reaper himself a keen fisherman (an idea going back to Renaissance times), seen during a rare holiday tying up a black mayfly pattern.

Part of the attraction of mayfly season is its highly visible mass migration – a dramatic eclosion that is soon swallowed up (by hirundines and others) into the food chain, as aerial plankton. Fishing often concerns cycles of life and death. A 'blizzard hatch' seems to energise the whole waterside, not least the human population. In Hampshire it causes a hatch of Range Rovers, and on the great limestone loughs of western Ireland – where I have never yet hit it right – the annual 'burst' still affects local communities for the best part of a month. Dapping persists here, and village gossoons would gather the live 'drakes' and sell them to anglers heading for their boats. Traditionally, the method was to mount two naturals on the blow-line, with maybe a sprig of gorse added for visibility, then wait as some great fish arced at your fly like the bow of Ulysses, not setting the hook until you had muttered, 'Ave Maria'. Then, come lunchtime, there would be an island fire of arbutus wood, the chortle of Kelly Kettles and maybe a sharpener of Powers whiskey from your metal cup. In his sometimes overly fanciful mythological study *The White Goddess* (1948), the visionary poet Robert Graves assures us that lough trout make a kind of 'dry squeak' as they take a mayfly, and that an erotic spring dance once celebrated the Irish princess Dechtire, who conceived her son – the warrior hero Cúchulainn – when she swallowed a mayfly on her wedding day, so 'he was able to swim like a trout as soon as born'. This truly is an inspirational insect.

The dragonish, silt-burrowing nymph may have been living in the substrate for more than a year when it responds to certain photoperiods and, in its final instar, rises to the surface, splits its cuticle and emerges into the limelight as a sexually immature green drake, or dun. It has a body the colour of buttermilk, and sports gauzy wings. Fish grab them like popcorn, though the only one I tried tasted faintly of grass. (The author Vladimir Nabokov, also a distinguished lepidopterist, once ate some Vermont butterflies and reported that the taste was 'like almonds and perhaps a green cheese combination'.) Unusually, the adult mayfly undergoes a second moult, becoming a spinner with long dark tails; you can see the apterous pellicles of the subimago left behind on leaves like rice-paper effigies (aw, shucks), as the now mature flies billow and jig aloft in a *danse d'amour*. The female has finally become a reproductive engine: her mouthparts atrophy, the intestine acts as an aerostat, and her abdomen is crammed up to the back of her head with as many as eight thousand eggs. During the mayfly mile-high club these are fertilised by males (once called the death drake), and then the spent gnat curtsies back along the water, depositing her precious burden in batches, until, splayed and exhausted, shivering in the surface film, she dies.

There are some thirty chalk streams in Normandy, and in the 1990s I visited the Risle, which I had read about in the writings of Charles Ritz, who loved the fishing at Aclou. In Brionne, the Doctor and I stopped to buy licences from M. Laroche, whose tackle shop was also a bar, so the paperwork was accompanied by a *pastis*, and we also purchased supplies of that notable pattern the Panama. It was mayfly time, and although there were opportunities for *nymphe à vue*, this was intended to be dry-fly sport.

During the day, the *truite fario* proved hyper-selective, so we enjoyed leisurely luncheons on the bank – cider, and Pont-l'Évêque, and berries dipped in icing sugar. Our host, Bernard, had been a hero of the Resistance, once an inmate of Belsen (he had no fingernails), and his family house had been accidentally bombed by the RAF aiming for the railway line to Paris. It was hard, in that gentle countryside, to conjure past turbulence – it had that ambience of timeless serenity which was sometimes described as 'The Peace of the Edwardians'. I was reminded of the plangent description of certain streams in neighbouring Brittany just before the Great War, by artist

Romilly Fedden in his atmospheric book *Golden Days* (1919): 'Mother Earth is very near in those hours by the water-side, that are so long and golden.' Even at the time, I realised how precious those moments were. There was usually a spinner fall, and during the evening rise I often managed to coax up a couple of wildies – truly the Risle Thing.

Over the past twenty years, many of my chalkstream days have been spent with fellow members of a small roving syndicate known as the Bundha Club – an eclectic affiliation of six chaps, including a horologist, a vet, a television star, a captain of industry, a best-selling novelist, plus *moi* (several of us are authors – I think it no coincidence that chalk has an early association with writing). Though not as arcane as 'The Twelve True Fishermen', a group whose ritual cutlery is stolen in G. K. Chesterton's Father Brown story *The Queer Feet*, we do have our esoteric protocols and a complex system of unwritten rules, along with a catalogue of fineable offences that include humourlessness in the face of failure and, conversely, the over-efficient catching of fish. Lateness for breakfast is deemed inexcusable, and once, in my bleary rush to descend in time, I brushed my teeth with Savlon (fortunately not Dr Badger's quinine mixture). Unlike some associations, we are no Rod Squad intent upon piscicide. Despite the club toast – 'Good Health to Men, and Death to Fish' – we generally release anything we catch, and the emphasis is more on bankside badinage, and bottles of buttery Chardonnay (no coincidence, either, that some chalk makes a fine vinicultural *terroir* and cellarage). My father would have made an ideal Bundharian, I think.

As well as some of the classic blue-ribbon streams – Test, Itchen, Wiltshire Avon, Dun, Anton – we try each year to take several beats on the Dever, an exquisite bonsai chalk water that was a favourite of Halford's and offers some of the clearest possible peepholes into The World According to Trout. When permitted, it is one of my preferred venues for sight-fishing with the nymph, though given the clarity and shallow nature of the water you have to be subtle and circumspect about it, or the smaller 'messenger fish' will scoot upstream and warn the larger, target fish of your approach (*sceot* was an Old English word for trout). You also need to moderate your nymphing tackle, and resist the modern tendency to fish too heavy or big. While it's true that the 'plop' delivery may occasionally attract recently stocked fish, it is counterproductive on waters like this, where you are better off avoiding bulky bead-heads and trying tiny tungsten patterns that are as slender and heavy as Leonard Cohen lyrics.

I relish the three-dimensional aspect to sight-fishing, when you can concentrate on the reaction of the fish, looking for a shift in its position, or the slit of an opening mouth – what Skues, in a piece of doggerel, called the 'wee little wink under water'. You need to learn how to read their lips. Wherever possible, I am a purist to the extent that I begin with a freshwater shrimp imitation, which can be as tiny as a size 18 hook, but must have that crucial *Gammarus* shape, hunched like a defecating dog. For years, I model-made my own confection, with black eyes, mixed seal fur body, beige hackle and back of painstakingly applied Araldite, but now there are many easier synthetics and epoxy-type materials for achieving that translucent effect – I christened my pattern the Receptionist ('Hold the line, please, while I try to connect you'). Recently, I have taken to carrying an entire box full of flat-sided shrimps sculpted in Macedonia by the talented Igor Stancev. With the shrimp (and weighted bugs in the Czech or Polish style) I employ the 'induced take' flip of the rod tip pioneered by Oliver Kite, and in America known as the Leisenring Lift. It is an altogether subtler method than it first appears.

On one occasion I was invited to fish the historic Houghton Club's many miles of the middle Test. It was in nymphing season, and at the customary teatime changeover in their Stockbridge headquarters, prior to switching to the evening beats, I told a venerable member that I had enjoyed a most productive day. 'What did you catch 'em on, I wonder?' 'Polish nymphs, sir.' '*Polished* nymphs, you say?' came the profoundly suspicious reply. His Victorian forebears would have described me as 'a disgrace to the hickory'. At any rate, it was a decade before I was invited again.

As a Hampshire sexton once remarked to the eminent author William Senior, 'Any feller can hook a risin' fish; it takes a proper hangler to bring 'em uppards.' Sometimes you have to tease a trout into responding, but systematic and indiscriminate 'fishing the water' with a nymph will undoubtedly disturb the pools for those intent on the dry fly, so guidelines on its use seem sensible. I admit that there are times when you can successfully lay siege to a stubborn, unfeeding trout by repeatedly confronting it with a sunk nymph or bug, and some may consider this unsporting. I have dubbed this 'love bombing', after that electioneering ploy of identifying marginal voters and convincing them your candidate is sufficiently similar in appeal to the one they would normally vote for. In both instances, an unwitting target is being fooled into falling for a fake, and must suffer the consequences.

On a whistle-stop tour of four New Zealand lodges in 2018, I learned quickly about two nymphing methods that repay careful study. Although not strictly chalk streams, some of their spring creeks that run off pumice (such as the North Island's Taharua) are remarkably similar – except that in the background you hear the lawn-sprinkler crackle of cicadas in the aromatic eucalyptus. Truck and Trailer (or Hopper and Dropper, Klink and Dink) is a 'duo' technique with a nymph fished below a dry fly, usually attached on a dropper tied directly to its hook bend, the floating fly acting as both attractor and indicator. This worked well during a float trip down the mighty Tongariro. In the South, on the Mataura and Maruia rivers, I was also tutored in the use of yarn strike indicators above minuscule tungsten nymphs on long leaders – a world away from our own chalk spider, but well worth the 23,000-mile round trip.

Modern European competition methods have devised several ingenious ways of detecting a subsurface take when your fish cannot be seen – watching a long French leader, or using a zero-weight fly-line – but sometimes you just have to rely on instinct and a sort of sixth sense, which comes with practice and watercraft. That is something you have to learn, but which can't easily be taught.

For a while I had a fortnightly rod at Compton, one of the more attractively managed estates on the Test below Stockbridge, with three beats and some delightful carriers. In a hatch pool by the hut on the Middle beat there was a roped-off area where several burly fish churned insouciantly around in a no-fly zone, and it occurred to me that, if one arrived a little earlier than the usual ten o'clock rendezvous with fellow members, one might ambush these huge specimens in the brisk run above the pool. So one Friday I parked well back, slid along the grass like a serpent (I was practically in Paradise, after all), peeked through the vegetation into the stream, and there were two brownies feathering the current with their pectorals, happily gulping some subsurface items. A short catapult cast sent my Receptionist into the zone, and the smaller of the brace swayed over and had it. Setting up on him, but still prone, I thrust my already extended landing net into the stream below, and as he turned tail for his downstream sanctuary he swam headlong into its meshes. From soup to nuts, it had taken perhaps two minutes. He was as fit as a butcher's dog, and weighed 11 pounds and 11 ounces – my largest brown trout, by a long chalk. Does it really count, since subterfuge had waylaid what was clearly one of the keeper's special protégés? Damn straight, it does.

Chalk-fed lakes can prove useful laboratories for developing fly tactics, and during the early 1980s, when I was a postgraduate student in London, I was a regular rod at Latimer Park Fishery, through which runs the Buckinghamshire Chess. From one of the little boats, with the right light, you could see good-sized trout cruising in clear depths of the bigger lake, which was always instructive. I did quite well there with a weighted pattern that had a body of wound dental dam, and also with flies like Richard Walker's Chomper, and marabou Damsel nymphs. These apprenticeship years coincided with the proliferation of small still-water venues – Damerham and Allen's Farm in Hampshire being two of the least artificial in ambience – and I probably learned more in those places than I ever did researching Augustan verse satire in the old Reading Room of the British Library.

Not that far from Central London is a wonderfully preserved private lake, into which flows a chalk stream that is one of the very few places where rainbow trout breed over here in the wild. I was fortunate enough to be invited there quite regularly, and I suppose it remains the most exciting trout venue I have ever visited in Britain. Lightly fished, with invariably clear water, it afforded you a gentle day paddling the little boat towards clearings in the weed patches, where a slowly worked nymph frequently created a miniature Charybdis of a take. Fred Buller was often there, and on one occasion we had a shore lunch cooked for us by the legendary specimen hunter and author Fred J. Taylor, which involved three hours of hedgerow wine and campfire songs before the old Desert Rat stripped off and dived headlong from the rickety landing stage. We finished that day afloat in the moonlight.

On 1 July 2002, I was being rowed there by my generous host along the mouth of the stream, figure-of-eight retrieving an Angora-bodied stonefly pattern off my 4-weight floating line, when a bow wave homed in from under the branches and I was taken by a very large rainbow. He muscled his way into some pond weed, but at length I coaxed him out, blindfolded by salad. It was a curiously hump-backed specimen – a slight deformity possibly caused by indigestible crayfish parts – and he weighed 7 pounds 4 ounces: the record for the lake, and an absolute trophy for a wild rainbow here in England.

If I ever do better than that, I doubt I would believe the story myself.

Thanks to my late friend John Hotchkiss, whose rich, boeuf bourguignon voice had graced the West End stage before he became a fishing guide, I have for some years now enjoyed membership of a club that controls several lovely beats of the Itchen. John (in Walton's phrase, 'An excellent Angler, and now with God') was, like me, something of a 'nymphermaniac', and once helped me achieve an 'Itchen Slam' – perch, pike, rainbow, brown and grayling all in one day. It was the pursuit of grayling that especially enthused us.

I have caught them from the Tamar in the west, to Teviot and Tay further north, but mostly I like to hunt them down in our clear southern waters around Halloween, when the bankside sward is gripped with frost, the brown trout are pairing off as if it were freshers' week, and low orange sunlight slanting through the branches is ribbed with bonfire smoke. I generally begin with a mug of Pirate's Blood (hot Bovril laced with dark rum), but it's a short day so you must keep going and eschew the temptations of the local saloon bar's welcoming taproom.

The spectral grayling used to be called an umber – this may derive from the term for the shadow cast by a sundial's style – and its Linnaean name *Thymallus thymallus* supposedly refers to the aroma of thyme some detect upon its skin (Cholmondeley-Pennell reckoned the umber was redolent of cucumber). With a cute pointy snout, underslung mouth and an eye that holds its indigo pupil within a teardrop of gold, the grayling is a subtle-looking fish sporting flanks of a pewtery subfusc laminated with nacreous bloom. This has lent its name to a type of wintry skyscape. Its trademark dorsal, like a blotched butterfly wing singed with madder along its cusp, is a thing of wonder: the Esquimaux, who harvest its Arctic cousin for dog fodder, call it *hewlock powak* – the fish with a fin like a wing. In 1867, Francis Francis christened her 'Lady of the Streams', but the German word is *Asche*, denoting her cindery hues, and I like to think of the elegant grayling as a Cinderella fish, pale and fastidious, with touches of rouge, a fluttering fan, and that captivating eye. Loire legend maintained that the grayling fed on nuggets of gold.

Yet not so long ago they were despised, and culled as vermin from chalk streams for the capital crime of supposedly eating trout eggs. Otherwise-sane pundits like Skues and Sheringham reviled them, though the ever liberal Francis condemned any angler who killed grayling out of season, hoping that on your return home 'your wife will have locked up the brandy, and gone out for a day or two'. *Thymallus* may have an adipose fin (that heraldic badge of the salmonid clan), yet it spawns in springtime along with coarse fish, so it was regarded as having been born on the wrong side of the piscine blanket and enjoyed a limbo status among certain sportsmen – Patrick Chalmers likened it to some Victorian country governess, socially not quite belonging upstairs or downstairs. More recently, it has been rehabilitated, and even become something of a cult – our 'Fourth Game Fish'. Cinderella may have come late to the ball, but many now consider her the belle.

Particularly when concentrated into schools of yearlings, pinks or shots, the grayling can prove most obliging, and less inclined than trout to scatter in the face of stranger danger. Provided you do not loom over them, several may be extracted from a shoal before they become gun shy – in a couple of Itchen competitions I have managed a basket of more than forty (a Tenkara rig can be ideal). On the freestone rivers of Yorkshire, there is delicate sport to be had long-trotting a red worm off your centrepin (Arthur Ransome described the 'thin laughter' of ice forming between your waders and the bank), but in the southern chalk I mostly take them on bugs as they rootle around the substrate, the solitary cock grayling sauntering up the gravel like a boulevardier stopping occasionally to take a pinch of snuff. They will readily feed at the surface, though, even in a snow flurry. The rise is abrupt, and your response must be as fast as chain lightning (Ransome's father advised him to strike 'a quarter of a second before' the fish reached the fly). I favour very small dries, like a Red Tag from the Welsh Borders, or that Ure pattern Sturdy's Fancy, though a modern Pinkhammer can be lethal. Our UK record stands at over 4 pounds; the ultimate Holy Grayling would weigh 5.

It used to be said that while the count of the Rhine dined off trout, the German emperor preferred grayling. I don't much care for them to eat (these days if I do kill one it's for pike bait), but some like them grilled with field mushrooms and elderflower wine, and Kite recommended a mustard sauce. To the proper chalkstream 'hangler', she is, at the onset of winter, one of the river's silver linings.

11

OF COLLECTING

'Mu', mu', mu'.' ('Words, words, words.')

**William Shakespeare, Nick Nicholas and Andrew
Strader, *The Klingon Hamlet* (2000)**

Anglers tend to share a collecting gene – after all, our passion is for collecting fish: it's the gathering part of the 'hunter-gatherer' pedigree and it accounts for much of the pleasure you can have with a penchant for fishing when unavoidably away from the water.

Proprietors of tackle shops know this. They offer a sense of plenty and a subtle prospect of the undiscovered, where goods are proverbially displayed first to catch your hungry eye – those multiple compartments of flies, for instance, among which you are convinced lies a Tegucigalpa Special that will prove your magic bullet. It's like a scribbler searching for the *mot juste*. Then you pass the stands of vertical, weapons-grade carbon: it amuses me to watch punters sagely waggling a new rod, because that way you can tell very little about its action (it's a salesman's ploy). Ask instead to rest the tip of the rod, Zorro-like, on his Adam's apple while he delivers his patter, then feel if the vibes are transmitted well and true (this won't work if your fly shop is run by a lady, but in that case you are anyway one lucky sonofabitch).

The trouble is, there's always something new – one of the earliest patents was granted (in 1632) to a Thomas Grant, for 'a fish call, or looking glasse, to lure fishes to nets, spears and hooks' – and I have never walked out of my favourite tackle emporia without acquiring some unneeded doodad, gizmo or gimmick. From a Bahamian beach shack I bought my first Billy Pate reel; the Farlows tented outpost on the banks of Russia's Rynda furnished me with yet another badged fleecy; from W. W. Doak, of Doaktown on the Miramichi, I lugged away several boxes full of Bombers, Popsicles and Green Machines; in the badlands of Miami, where my cabbie refused to wait, Captain Harry's sold me a dozen Surgeon's Tube lures for barracuda. Walton described the cynic Diogenes survey-ing the wares in a country fair, which ranged from fiddles to nutcrackers, 'and, having observed them and all the other finnimbruns, he said to his friend, "Lord, how many things are there in the world of which Diogenes hath no need".' The crusty old Athenian would have made a lousy angler.

The ideal tackle shop is sometimes likened to Aladdin's

Cave, and I've heard that said about my own rod room, though I prefer to think of it as a cabinet of curiosities – those *Wunderkammern* that were so fashionable in the Renaissance, chambers filled with extraordinary objects, organic and inorganic. One of the greatest was assembled by the alchemist Ole Worm (now, he *must* have been a fisherman), and was the first such collection ever to be called a 'museum', or place of the Muses, later rehoused in Copenhagen Castle. I am not really a proper curator, since I am neither systematic nor unusually knowledgeable; I seldom categorically seek out rarities, and have a limited grasp of taxonomy. I suppose I am a sort of maximalist and merely indulge my manifold magpie tendencies, and collect carefully but unpredictably, like any child. I am a gearhead, a tackle tart, and a catholic accumulator of all types of piscatoriana, kit, caboodle and clobber. *'You don't need all this stuff,'* insists my civilian spouse, as yet another desirable item is installed in what used to be the kitchen of our Perthshire home. *'That's just feeding the monster.'* She is such a dedicated de-clutterer that once she disposed of my lifetime's collection of vinyl records when I was away fishing for peacock bass in Venezuela. *Ay, caramba!*

Crammed with desiderata as it is (*'Rubbish, tat!'*), my tackle-room retreat makes a perfect workshop during the silvery light of a winter's afternoon wherein to perform what Ransome called 'small varnishings' – items of maintenance such as the lubing of reels, waxing of zips, and testing of rod rings for nicks (the latter is achieved by running a silk stocking through them). On the felt-lined racks along one wall there is a horizontal multiplicity of rods, though my collection actually numbers ninety-one, so many others are stacked in their arsenal of launch tubes in one corner. The spectrum comprises a 1-weight toothpick model suited to 'flea' fishing and a 14-weight used for hurling roadkill patterns at trevally. There are currently 121 reels, ranging from a serious Bogdan (why settle for lambswool when you can buy cashmere?) to a Shimano Stella fixed-spool, capable of yanking a jetskiing bay turkey over on his back, though I am also fond of the old-fangled, and possess some early wooden pirns. The reel – probably a spin-off from the Sung dynasty sericulture bobbin – is not mentioned in Britain until 1651, when Thomas Barker describes a basic barrel winder, but the introduction of a running line (as opposed to one merely affixed to the rod tip) revolutionised the sport, and I find in the grace notes, ingenuity and variety of my examples an aesthetic delight that derives from their 'beautility'.

I have somehow amassed eight pairs of waders, ten fishing vests (that admirable invention of the American maestro Lee Wulff), 118 floats (including a Snoopy Bobber, and a specimen made from moose droppings), thirteen pairs of polarised glasses, two cupboards and four drawers stacked with fly boxes, and, strung beneath a shelf on many dozens of plastic spools, from Skagit to Slime, a line of lines. There is a little memorial cairn of stones from hallowed waters I have visited (Redmire Pool, the Alta), and a pinboard of honour where I have retired battle-scarred veteran flies, each with its peculiar tale to tell. I suppose that all of the above offers a snapshot of an obsession. Then we come to what I call the inimitable 'etceteras of angling'.

Glass minnow traps, smelly jelly, mystic bloodworm paste, bloody tuna secret sauce, Swedish fish candy, live hopper harness, Roman fish hooks, crappie jigs, bait kettles, alarm clocks ('Fish on, man!' they screech, at oh-daft-thirty), immense exhibition plug baits, a carved and painted permit, a taxidermised kingfisher, a carved carp, skulls from crocodile and hutia, a shark's-head stand for my clutch of wading sticks, a stool from Walton's parish church, five iterations of Billy the Singing Bass (batteries removed, by order of The Management), myriad bottles, watches, boxes, tins, rugs, posters, plates, cruet sets and food packages all bearing pisciform designs, an RAF survival hook and handline outfit, wicker creels, Sammy the stuffed Fishing Stoat (complete with miniature Tonkin cane rod and sombrero), a Canadian fish bonker plus an Alaskan wife tamer (both priests, as this is a place of sanctuary), twenty-three assorted glaives, gaffs, leisters, harpoons, bidents and tridents, none of which would disgrace Neptune or Britannia, a South African marlin lure modelled on an erect human penis (XL), Flyagra floatant, two cabinets of vintage spinning baits, and a bath plug attached to a Japanese sea float which, if you suspend it over the outflow, will be tugged and nibbled by the Coriolis force of the water, just like an eager fish ('*I believe you need help*').

Some of these bibelots and *objets de vertu* ('*Dross, dreck*') have histories that invest them with the aura of reliquaries – if my collection is a shrine to the cult of fishing, it is only in the original sense of religion being something based on a binding together, the making of connections (that central act of faith on which all angling is based). But with relics one has to beware of imitations – the arm of St Anthony in Geneva turned out to be a stag's penis bone. The cultural history of some of my more treasured objects is more numinous. I have the avuncular Admiral, of course, and the Doctor's father's venerable centre-pin reel, flies from the collection of Richard Waddington and Hugh Falkus, and two boxes that belonged to the Irish baritone Harry Plunket Greene, author of that enduring hymn to the Bourne, *Where the Bright Waters Meet* (1924). When I handle these things, my imagination summons their original owners – in the singer's case, his village community before the Great War, described as snappily as in any Saki story, with its cast that included Sir Seymour Sharkey, Moneypenny of 'The Thunderer', and the little stream that he saw dramatically decline. My admiration took me to the graveyard in Hurstbourne Priors, where I knotted an Iron Blue around his gravestone.

I am still scouring auction catalogues for John the Baptist's wading staff.

Ray Kroc, supreme ruler of the McDonald's burger empire, once questioned whether it was really any stranger to take delight in the texture and carved silhouette of a bun than to admire the hackles on a fishing fly. He realised that some fly-fishers can become fixated about rolling their own (Klinkburger, anyone, Quarter Templedog with Cheese?)

and he must also have had a well-developed feel for various types of devotion, because when his widow died she left £820 million to the Salvation Army. What he is describing is feathermania. In the window of my rod room is a dedicated fly-dressing station glistering with fine-drawn wires and tinsels, plumed like the boudoir of some superannuated chorus girl – toucan, marabou stork, jungle cock, macaw, kingfisher, Indian crow, vulturine guineafowl, florican bustard, Amherst pheasant, ptarmigan, along with hair or fur from spaniel and mole, collie, peccary and Arctic fox. There are stacks of bead pots from craft shops, glue guns, rubbery prosthetic shrimp legs and synthetic mayfly bodies shaped like a mouse's condom, a foam rack impaled with salvaged surgical instruments (scalpels, tweezers), three vices (including a travelling set fashioned by Dutch jeweller Ari 't Hart), hanks of artificial winging fibres and, above the serried hook caddies, a seemingly empty package of Partridge Invisible NO/C1 hooks, sent out as an April Fool's promotion.

So much depends on your hook, and the search for perfection continues. Back in Walton's day (when hooks were often modified needles or nails), the ideal was thought to be shaped like the hoof-print of Pegasus. Tempering processes improved with the Industrial Revolution and hooks reached a wide colonial market, being used as currency, jewellery and even for the cleaning of toenails. In Oceania, a ship's nail was so valuable for bending hamiform and converting into fish hooks that visiting sailors could trade one for the amorous affections of a local island lassie (or laddie, perhaps). G. S. Marryat reckoned a hook's qualities should combine 'the temper of an angel with the penetration of a prophet – fine enough to be invisible, and strong enough to kill a bull in a ten-acre field'. One day perhaps we will have a near-invisible, transparent plastic hook, and there will be no need for such April Fools.

Fly-fishing must surely predate by ages its first surviving mention in writing, from the second century AD, and unlike those who pursue other sports anglers have often had to construct their own equipment (not many enthusiasts carve ice pucks or hand-stitch rugby balls). Ingredients for fly-dressing have long made up part of an angler's kit list. The arch foe of Walton, Captain Franck, had this to recommend in 1694 to any reader embarking on a tour of Caledonia: 'Always carry your dubbing-bag about you, wherein there ought to be silks of all sorts, threads, thrums, moccado-ends [mohair] and cruels of all sizes, diversified and stained wool, with dogs and bears hair, besides twisted fine threads of gold and silver, with feathers from the capon, partridge, peacock, pheasant, mallard, smith [a fancy pigeon], teal, snite [snipe], parrot, heronshaw, paraketta, bittern, hobby, phlimingo, or Indian-flush; but the mockaw, without exception, gives flames of life to the hackle.' I do have some phlimingo, gleaned from the crater at Ngorongoro, but the bittern is becoming booming hard to find.

In the Victorian era there was a sudden vogue for using gaudy and elaborate flies for salmon. The impetus came from Ireland and spread among sportsmen heading to the newly fashionable estates of Scotland, where some of the staider gillies claimed the new patterns were scaring the salmon away. This craving for exotica soon had milliners and aviarists scrabbling to keep up with demand for ever-more outré materials – some of the more recherché patterns required yak, leopard or lilac-breasted roller – and this coincided with the female mode of feathery 'picture hats' that threatened to decimate certain species. These included the bird of paradise, which had to be killed in its full breeding dress. This fetish for novelty reached its apotheosis in 1895 with the publishing sensation that was *The Salmon Fly* by George Kelson, an ardent self-publicist who angled in a bowler hat treated with Acme Black, and stockings specially commissioned from the workshops of Dublin's jail. Some of the barbed beauties he catalogued were devised by his chum Major John Popkin Traherne (a military man with a racy private life), and the list of patterns sounds like a rush of Regency racehorses: Bonne Bouche, La Gitana, Dusty Miller, Tom Tickler, The Nepenthian. Eventually, this rage for imperial showiness waned, around the time battle-dress scarlet gave way to khaki in the field, and now only those who tie exhibition pieces are likely to know how to build a fly that might involve a hundred sequential procedures.

One casualty of this heady period was Edward Fitzgibbon, a successful author and journalist who wrote under the nom de plume 'Ephemera'. He was banished to Greece by his family for overstepping the mark of propriety and incorporating into some of his flies hair cropped from the intimate areas of his ladylove (he perished of drink, aged fifty-four).

The practice continues, as I have mentioned regarding my friend the Laureate, but that's enough about fishing-related finnimbruns.

My tackle haven should perhaps be seen in the context of my other collecting impulses, one of which is for books (all texts are themselves but collections of 'words, words' – essentially dishevelled dictionaries). My library stops one click short of bibliomania, but the room houses evidence of other agglomerative tendencies and a weakness for miscellanea (*'junk'*). Above the fireplace is a wall collage that combines some 350 assorted cigarette packets, representing one-third of a collection I made in the 1970s, most of which I smoked during my gap-year travel or student days (the centrepiece is a scarce carton from Air Force One, obtained for me by Benazir Bhutto). Along the mantelpiece is arrayed a selection of *bondieuserie* (*'tourist geegaws'*), including small icons, Mexican Day of the Dead figurines, votive candles, rosaries (one given to my father by the Pope during the Italian campaign) and Lourdes water in bottles shaped like the BVM – of curious appeal, even to an atheist. I have never much wanted to own cased specimens caught by other anglers, but beside my club fender sits an extended family of stuffed kittens discovered in various Greek islands, part of a modest taxidermal menagerie from cobra to capercaillie that adorns the house, featuring a baby *sanglier*, a Thai bat, a chameleon, a Parisian fighting dog's mounted head, two piranhas, an armadillo lamp, red squirrels, ducklings, owls and several stuffed rats, one of which is attired like Winston Churchill (I confess that item is a tad tonto).

Elsewhere you may encounter a shelf displaying numerous Marmite jars (with delicious limited editions), an alcove devoted to salesmen's selections of false teeth and glass eyeballs (some from the American Civil War battlefields), and a fine spectrum of Noxzema shaving foam canisters (I like lines of things, and variations on a theme). Learned analysts might term these 'compensatory objects', but I couldn't give a Donald Duck, as I believe even the mundane can appear monumental, working on the premise that under certain circumstances one plus one can make eleven. My vitrine in the dining room contains an arrangement of 132 different packaged soaps – a disposable and near-global product with an infinite series of incarnations, mine embracing (apart from carnation itself) Swedish pine, Dead Sea salt, French absinthe, Portuguese magnolia, Greek donkey milk, Bora Bora frangipani and Icelandic volcano ash. If I had to single out one, it would be Profumo Classico, with its classic black and gold box, and fragrance of blackcurrant with mint. Well, I would, wouldn't I?

I have for decades had a fondness for lumber, kitsch and collage, combining *objets trouvés* discovered on bric-a-brac stalls from Alaska to Tibet, either given as presents (I am not hard to shop for at Christmas), or trophies personally found and liberated – no internet shortcuts allowed. While I am not claiming for myself anything loftier than a little kindergarten surrealism, I do enjoy the work of constructivist artists like

Kurt Schwitters, whose original multi-room *Merzbau* instal-
lation in Hanover was destroyed by Allied bombers, though
as an internee on the Isle of Man he continued to make sculptures out of porridge. I am also fascinated by the abstract artist
Joseph Beuys, who, wearing a fishing waistcoat, would lecture
with his face covered in honey, cradling a dead hare, alleging
his *oeuvres* (which included a *Fettecke* carved from 5 kilograms
of butter) were inspired by his experience of being swathed in
fat and felt when his life was saved by nomads who rescued
him after his Stuka was shot down over Crimea. I also admire
Marcel Duchamp's readymades, and though I have no urinal I
do possess a miniature lavatory, an exquisitely bad porcelain
souvenir from Portofino inscribed with the incontrovertible
sentiment *Tutto il saper del cuoco finisce in questo loco* (all the
expertise of the cook ends up here). This is convenient, as
some regard my passion as symptoms of an anal retentive, or
at least a hoarder suffering from 'disposophobia', but, as I try
to explain to bemused house guests clutching their cocktails,
by realigning and juxtaposing various bits and pieces I am
in fact attempting to counteract the effects of an increasingly
fragmented world.

You may well consider this all hokum and horse feathers,
lector benevole (as does She Who Sold My Captain Beefheart
Albums). The Freudian analyst Wilhelm Stekel (*Disguises of
Love*, 1922) reckoned collections resembled harems and were
a symptom of love hunger and psychic restlessness. This may
be true of someone who dreams of philately, or exclusively
gathers Camembert cheese labels (a tyrosemiophile), but I

do maintain that collecting, as another form of connecting, is central to the act of angling, which involves threading experiences together in your own idiosyncratic way. I can't help but wonder if this is all part of our (restless) search for a coherence which was lost when that flaming sword at the east of Eden ensured re-entry to Paradise was forever denied us.

The list is one of the most ancient shapes of the written word: from the Old Testament (with its rhythmic Genesis 'begats', Commandments and Beatitudes) to epic farragoes, shopping lists, taxonomies and checklists, recipes, wills, legal codes, inventories, menus, accounts, resolutions, book indexes. The latter offer core samples of a work, a strikingly fine example being the volume-long index to Robert Latham's magisterial 1983 edition of *The Diary of Samuel Pepys*, with its piquant entries on Sex Life (and a tantalising allusion to 'paradise fish'). As Eric Griffiths astutely explained in his posthumously published Cambridge lectures, *If Not Critical* (2018), lists can be potent as well as trivial since word order may reveal, in its series and sequence, internal patterns of resonance that make them more than the sum of their parts. They can offer a kind of pied beauty, and often relate to the making of plans. It is hardly surprising that the literature of fishing is well stocked with them.

Arthur Ransome recalled an old angler who ended his pre-departure checklist with 'Kiss wife'. The tackle list is a staple of this genre, and I like William Gilbert's advice (from 1676): 'Fail not to have with you, viz. A neat rod of about four

Foot long, in several pieces, one with another. Two or three Lines fitted up, of all sorts. Spare hooks, links, floats, wax, plummets, caps etc. And if you have a boy to go along with you, a good Neats-tongue, and a bottle of Canary should not be wanting.' For his 1927 safari to New Zealand aboard *The Fisherman*, the bombastic author Zane (christened Pearl) Grey scrupulously itemised an entire page of tackle requirements for the pursuit of big game species, including 161 rods, fifty-one reels, gigs, self-striking spoons, nine dozen gaffs, 500 pounds of sinkers, two hundred lines, and 10,000 feet of rope in tubs for his harpoons.

Grey logged his catches meticulously – billfish, tuna, sharks – and the personal record book is another type of piscatorial list that enjoys a mosaic history. I have kept a note of practically everything I have caught since the age of twelve (and entries for certain unproductive days that were still in some way noteworthy) and it is remarkable how, even at a distance of decades, a few salient details can serve to rehydrate the experience of time spent long ago on the water. My four volumes are impressionistic and uneven, with miscellaneous notes about the context of a particular day on the water – a bycatch, perhaps (these have included swallows, ducks, a dog, a cow and several guides), or the odd aside ('Bob the keeper claimed to have been a cousin of Marlene Dietrich'), along with numerous regrets, waterlogged blanks and missed opportunities. Memory Lane is thickly brambled in places. Ransome added, 'The habit of noting such things strengthens the habit of observing them,' and felt his diary was essentially 'writing a letter to himself when old'. He anticipated one of angling's greatest pleasures, that of re-collecting.

More methodical records can become statistical ledgers now of historical value, and one is frequently astonished by what our forefathers could achieve. Cyril Wells (an Eton housemaster, who had twice bowled out W. G. Grace) killed Norwegian salmon of every weight between 20 and 58 pounds, missing only one of 55; Robert Pashley, fishing the Wye from the 1890s for fifty years accounted for ten thousand salmon of which twenty-nine were 'portmanteau' specimens weighing 40 pounds or more; and I savour the nonchalance with which Mr Pryor entered his catch on Upper Floors, October 27th 1886 – fifteen salmon, 'Heaviest 57 pounds.' The narrative in one Hebridean lodge book reads: 'August 14th. 1914. War declared on Germany. Wind went South in the afternoon.'

An entire anthology could be made of outlandish baits, pastes and decoctions that have been devised down the years. Some early manuals read like grimoires, complete with actual spells and alchemical recipes: a monastic manuscript from Heidelberg recommends carp bait compounded of beaver testicles admixed with barley and human blood, while another insists on the efficacy of fox lungs, or a vulture's foot. The cook and author Thomas Barker passed on a confection from the Sun King's personal apothecary-royal: 'take cat's fat, heron's fat, and the best assafoetida [an aromatic Asian gum], mummy finely powdered, cummin seed, two scruples, and camphor, galbanum, and Venice turpentine, of each one dram, and civet [musky perfume] two grains, which have to be glazed in a gallipot for two years, kept in a pewter box, then used to anoint hook and line. *'Probatum est,'* he reassures us. Modern-day carp specialists – their secret 'boilie' formulae laced with essence of krill, chicken tikka, or banoffee pie – are surely direct heirs to this tradition.

There is a similar sense of appreciation to be had from a colourful compendium of fly names, the pastime's sumptuous dramatis personae that is literally a cast list. Queen of the Waters, Ice Maiden, Satan's Monkey, Dusty Miller, Spring Rain, Parmachene Belle, Tup's Indispensable, Fairy Sedge, Night Hawk, Silver Tiger – just reeling off their names can fire one's enthusiasm. How flies come to be christened is an intriguing business. The Goat's Toe is tied with nothing goatish (it's an Irish phrase denoting something foolish); the Soldier Palmer refers to an early caterpillar style of dressing, after Crusaders returning from the Holy Land bearing palm leaves; the deadly prawn imitation Red Frances was named after the secretary of its deviser, Peter Deane; the venerable Coachman commemorates Tom Bosworth, coachman to William IV, who was so adept with his whip he could dash the pipe out of a passer-by's mouth. There are eponyms we can confidently trace – Peter Ross was a Perthshire shopkeeper, Willie Gunn a Brora gillie – but few now recall the models for Hamlin's Eyeball, Miss Julie, Sir Sam Darling, Mrs Haase, or Rat-faced McDougal, all once considered bywords for success (the latter was originally named the Beaverkill Bastard).

Some of the more modern patterns can sound a little less

lyrical – Bitch Creek Crawler, Madam X, Woolly Bugger, Psycho Puff, Chernobyl Crab. Bostonian cabbie Jack Gartside (the William Burroughs of fly-dressing) designed a saltwater pattern called the Turd, which does slightly resemble a mini-version of its namesake, though it's doubtful that's what bonefish take it for. There is also a crab pattern popular for permit imaginatively dubbed the Merkin, after an eighteenth-century genital toupee for unfortunate women depilated as a consequence of mercury treatment for venereal disease. Merkin was also the first name of President Muffley, one of the roles played by Peter Sellers in the 1964 movie *Dr. Strangelove*.

Lists of different knots and their stages of construction are many centuries old, the earliest English example in print dating from 1496, for the 'water knotte or elles duchys knotte', still widely in use today, though it's unclear which duchess we have to thank. Knots have had many associations with magic and witchcraft (along with seafaring, of course) and to the angler they are a common denominator without which that line of connection cannot be secured to your quarry; you have to be self-sufficient in tying them, and I feel one should always be safe and pack one's own parachute, even though some of the more ingenious ones are of almost Gordian complexity (I am speaking here of deliberately created knots, rather than tangles, fankles or bourachs). Palomar, Bimini Twist, Blood Bight, Egg Loop – there is a certain music in these names, too, maybe the sound of indie bands: Paragum, Surgeon's Loop, Improved Turle, Allbright, Perfection. The knot is central to any angling narrative, a crucial copula that links fish and fisherman along the lightning thread.

One of the largest collections of terms in the fishing lexicon embraces the names of pools – or 'casts', as once they were known in Scotland. Their very mention can act as shorthand between two anglers, invoking a shared memory of the sweet rattle of currents, the gleam of spray against boulders. Pool names can be like old friends – we each have our personal favourites on mental speed dial, their highlights as familiar in memory as the features of your beloved. Some are more memorable than others: Bridge, Long, Junction and Hut are probably the most common, whereas certain Icelandic designations can defeat the average visitor (on the Laxá í Leirársveit in Iceland we had to resort to pedestrian numbering, so that the Narfakotsstrengur became simply 24-A). The longest place name in the United States, a lake near Webster, Massachusetts, means in Nipmuc 'the neutral boundary where nobody actually fishes', and is easily discussed among fellow sportsmen as Chargoggagoggmanchauggagoggchaubunagungamaugg.

Some are distinctly unglamorous (Car Park, on the Alness; Sewage, on the Nith) but others comprise a litany of evocative toponyms – Mausoleum, Rumbling Stane, Upper Smashie, Back o' the Bog, Madman's, Devil's Chair, Maharajah's, Burn of Angels. Income Stream, on Floors sounds lucrative, and Whiskey (the Doon) most refreshing; there are the Rocks of Solitude (North Esk), Grief (the Shin), a Vale of Tears (the Helmsdale), Suicide (the Scottish Dee) and a Deadman's Hole on the Torridge. In 1993, on a visit to his home at Court Green,

I was once trying to wax poetical with Ted Hughes as we walked the Torridge and at one pool where some concrete sections had been introduced to boost the flow I asked, in my best, breathy acolyte's tones, 'So, what do you call *this* pool?' 'Concrete,' came the stony response. 'We call it: Concrete.'

It is a signal accolade to have a pool named after you. On an exploratory trip in 1991 to the Kola Peninsula's Middle Umba, we missed this opportunity by just a week. The previous party had already tagged all the likely runs, one being known as Captain's Hat, where an SAS officer's titfer had been blown off. There is still no Profumo's Stream. I have fished Willie's Pot, Lord March and Mrs Coke's in full knowledge of their origin, and it's clear (from their profiles) why the Gruinard has two rocky landmarks nicknamed Hitler and Mussolini; Blacksboat, on the Spey (at Ballindalloch), is named for a Laird's black servant who used to operate the ferry there, but what is the secret of Sam's Box (Orchy), and who was the limping lady of Makerstoun (on Tweed) who lent her sobriquet to Hirple Nelly? Only their names, like the rocks, remain.

I have mentioned my teenage triumph on Paradise (the Shin) and there's another heavenly place I know: Maison Dieu, on Teviot. There was once a hospice nearby, a lazar house that tended lepers when that disease was common in Scotland (Robert the Bruce was not afflicted with it, contrary to legend, though his father was), so I can truthfully say that I have taken a fish from the House of the Lord.

Bitterling, steelhead, rainbow runner, hickory shad, dragonet, alewife, snook, wahoo, brill, cuckoo wrasse – why should twitchers steal all the onomastic thunder, with their lovely cotingas and purple-sprouting spoonbills? Our names for fishes form a coruscating draught of monikers: the Gaelic for blood-red gives char its name, the pike is a luce (or shaft of light), the stargazer (*pesce prete* in Italy) has upward-turned, imprecating eyeballs, the porbeagle shark is a hybrid of beagle and porpoise, the humble sole was thought by Greeks to resemble a water-nymph's sandal. The bony horse mackerel is known in Provence as *estranglo bello-merò* (mother-in-law's garotte) probably for the ropelike pattern along its flanks. Some Linnaean terms sound quite apt: *Sprattus sprattus* is an ideal blue-collar name for that working stiff of the seas, and the charming *Rutilus rutilus* is the Name of the Roach, though for neatness nothing beats the technical appellation of a small Australian wasp known as *Aha ha*. The salmon is always, for superstitious reasons, simply A Fish.

Some were less fortunate during naming day in Eden – the Alabama chubsucker sounds like a sexual deviation, and *Halichoeres bivittatus* is known as slippery dick. With over thirty thousand species of fish being identified by collectors, there were bound to be some cumbersome christenings: sarcastic fringehead, midshipman toadfish, Iowan starhead topminnow, gaff-topside pompano, triplewart seadevil, convict surgeonfish and freckled madtom (I feel certain I was at the same Berkshire boarding school as him in the 1960s). The Dutch name for the famously phalliform lamprey – *beekprik* – has apparently lost little in translation.

'Names are not always what they seem,' warned Mark

Twain. 'The common Welsh name Bzjxxllwep is pronounced Jackson.' Thus, the Murray cod is a perch, the pike-perch is not a pike, neither freshwater nor marine dorado is a dolphin, and the rock salmon dished up in your chippie is a nurse-hound catshark (also known as bull huss, and greater spotted dogfish). As stocks of traditional table species are relentlessly overfished, what appetising new *noms de dégustation* is the food industry going to cook up to disguise megrim, lump-sucker and slimehead? Or will it just be a question of 'Let them eat Hake'?

Nomenclature can become a vexed issue. The curious John Dory may be named for its agreeable coloration (*jaune dorée*), or because it is thought to bear the imprimatur of St Peter's thumb when he caught one in search of tribute money from the Sea of Galilee – 'Cast an hook, and take up the fish that first cometh up' (Matthew 17) – the saint, as keeper of the keys to Heaven, is known as *janitore*. In Turkey it is known as *dülger* (carpenter) because its skeletal structure is supposed to resemble a set of tools. French trawlermen have settled for *l'horrible*. Its fancypants rockstar offspring ichthyological name is *Zeus faber*. So it goes.

Whichever quarry is on your list, always remember to 'Kiss wife'.

12

The Lightning Thread

'Good luck to the hand on the rod./ There is
thunder under his thumbs,/ Gold gut is a lightning
thread,/ His fiery reel sings off its flames.'

Dylan Thomas, 'Ballad of the Long-Legged Bait' (1941)

The Place of the Claw describes several high spur-like escarpments overlooking the Vale of Atholl, once the site of the chief of a sept of Clan MacIntosh; back in the tenth century, if he could walk dry-shod in times of drought to a certain rock in the nearby River Tilt, his fellow clansmen would propitiate their gods with a human sacrifice. In rainless summers the local saying runs, 'It's MacIntosh weather, right enough.' Our house was built up here a century ago, very close to the cartographical centre of Scotland, and thus within handy striking distance of much fishful Highland water: I do not own any river myself, but from my study windows I can see a lochan at the end of the lawn, and the occasional brown trout rising. Fortunately, this is spring-fed, so I never have to resort to pagan practices when sky water is in short supply.

I have fished some forty Scottish rivers or loch systems, but perhaps because at heart I am a trout man, its larger waters can seem daunting to try for salmon. Our local Tay – one of the Big Four – still offers the chance of a sizeable *salar*, but I am only ever happy on its higher beats. I have enjoyed some wonderful opportunities on the Spey, especially around Arndilly, but I have never come to know it well, unlike Tweed, and I always feel I am not quite covering it properly (perhaps because, when I was a schoolboy on Castle Grant, a grouchy gillie told me as much). The Dee at Dinnet and Cambus strikes me as the prettiest of them all, particularly in late spring when the air is redolent of pine resin and the banks bright with bluebells, and you can fish in your shirtsleeves after supper, throwing a full floater across some of the country's loveliest fly water. But I am drawn to the more intimate scale of rivers like the Kirkaig, Gruinard, Carron and Oykel, where you must keep well back, scramble into position, and read the water close up, work with the grain of its knots and whorls. I prefer to feel I am within a few yards of my fish when I'm hunting it.

The Tilt is a tributary of a tributary of a tributary of the Tay, and by the time the salmon reach this neck of the woods they are ready for a rest, and the prospect of some imminent lovemaking. With a late summer spate it rises rapidly, but is seldom dirty for long, as it runs off granite and limestone. Scrope, who stalked deer extensively around these head-waters, noted, 'this river speaks in a voice of thunder'. Queen Victoria used to admire its remote, vertiginous course when she rode over to Blair Castle from Balmoral. In 1907, the first ever British powered flights were trialled here in secret by the War Office, involving an aircraft modelled on a Zanonia leaf. The designer was J. W. Dunne, later known to anglers for his unorthodox book *Sunshine and the Dry Fly* (1924) but primarily distinguished as the philosopher who developed the notion of serial time, wherein past, present and future are deemed to co-exist simultaneously – not such a bad idea when spending the day on a water in these wildlands.

The Tilt was once renowned as a 'potchers' river, as there are many crannies and pots where (one imagines) salmon can be sniggled out or targeted with baits. It is typical of the sorts of stream where one has to be there at precisely the right time, and on just one occasion I really was.

A call from the castle said they were not using their house rods one August morning, and the pools were looking in good ply. I abandoned my desk, and within half an hour was crouching at the neck of Shepherd's pool, looping on a sinking head, and finding the right Garry tube for the height. A short switch cast covered the top run, and within minutes a slightly coloured cock fish snaffled it like a scaffolder's fry-up. Before I had reached the tail-out, three more salmon had snatched it from the white water, a brace of which made it to the bank, where my wife, periodically looking up from her paperback, presumed this was what salmon fishing was generally like, though the castle records suggested a similar feat had not been achieved for quite a few years. More often than not, of course, I return empty-handed from such rivers, because much of the time salmon seem to me to be frankly uncatchable.

Exactly why Atlantics ever take your lure has for ages been an enigma thudding around angling circles like a moth baffled inside a lampshade. I realise that, beyond some juice that resembles tobacco spit, hardly any evidence of ingested foodstuffs is found by biologists when examining stomach contents of salmon caught in fresh water, and I am aware that its body rids itself of digestive enzymes (which is one reason the flesh travels freshly so well), but I can't help persisting in the Flat Earth belief that *Salmo salar* does occasionally attempt to feed on its return to the supposedly natal river, and that when its hunger pangs are momentarily unsuppressed it tries to eat things much in the way a human succumbs to pica (gorging on paint chips or coal) and is perhaps responding to understandably complex hormonal influences, and experiencing cravings, like some pregnant women. I have only a hunch about this, as with so many other aspects of piscatorial theory schmeory, but I have learned to consider plausible almost any abstruse or crackpot explanation about the unpredictability of salmon, and I can think of nothing much else that would account for how I have caught them on natural prawn, shrimp, worm and sprat, leather eel-tail, wooden minnow, plastic plug, spun Swedish metal, dry fly, wet fly and dap, while others have taken fish on everything from carved carrots to helicopter ear plugs, and even painted bare hooks.

You can understand that a bunch of succulent lobworms might resemble the palpitating movement of a marine squid (few parr would ever have seen such a sight, though in later life a smell of the smaller worms juvenile fish do eat might reawaken an earlier river memory), and of course these are frequently swallowed. But what about the manifold variations in my cupboard full of fly boxes? The arguments are familiar – aggression, irritation, curiosity – and maybe, worked plausibly, certain patterns do faintly resemble a smudge of plankton, an attractive insect, an escaping capelin or streaking sand-eel that provokes a sudden reflex, a shard-flash of distant recognition, if you factor in various other influences such as barometric pressure, or moon and tide phases. But I, for one, do not wish for much further scientific input in the form of solunar tables, or watches that record atmospheric changes, and I am content to believe this unfathomable quarry is either feeding (in which case it was judicious of Mother Nature to make this a rare occurrence, otherwise there would be little left alive in the river) or else exercising sheer *joie de vivre* – playing for fun, as dolphins, octopuses or corvids seemingly do. (There are plenty of crackpot theories out there, and I like this as much as any.) These would be exceptional instances, because on many days you make a thousand casts that go unheeded. When salmon are stubbornly being salmon, attracting their attention is like trying to flag down a black cab that has turned off its 'for hire' light and is intent on heading back home. You can wave your arms around as vigorously as you like, clutching your Spey rod. Small wonder in City parlance a day of strenuous activity resulting in no deals was formerly dubbed a 'salmon day'.

There has been elaborate research, but we still know a scant amount about the mysteries of salmon behaviour. The pieces just don't fit together. A Hindu fable tells how six blind men were invited to describe an elephant: one feels its trunk and decides it must be a snake, another deduces from the texture of the knee that it is a tree, but individually they achieve no consensus. As a layman, my view of our most glamorous of migrants is somewhat similar (one encounters other sportsmen, of course, who apparently see the situation less equivocally): there is a modern version of the fable in which six elephants are bidden to describe by touch a man, and they all agree that he is flat.

A Gaelic phrase to describe something sudden that happens out of the blue runs is *Rug iasg orm*, 'a fish has caught me', and with salmon fishing I sometimes think the only explanation for these chance encounters is that the quarry has decided to catch *you* – a sentiment expressed by Ted Hughes in his poem 'After Moonless Midnight', when the river whispers of the sea-trouting angler, 'We've got him.' Fish can be so wonderfully mercurial. In July 2020, when the prolific Sheep Pool on Iceland's Haffjarðará was estimated (from clifftop surveillance) to be holding several hundred salmon, I fished it intensively for two hours without any response, and then quite suddenly, around seven in the evening, the air cooled and I landed eight fish in ninety minutes on a small Olive Frances – a feat I doubt I will ever repeat.

The variety of salmon takes is considerable, and I suppose it is this which gives it the edge for some anglers over other types of fishing – the mumbling of worms, that leaf-like brush at a sunk fly, the pronounced tunk that always astonishes you when it fails to take hold, a steady drawing away as you feed fly-line between your fingers into a slip-strike, the sudden onslaught on a spoon like some thunder blow from Thor's hammer. I am convinced fish frequently take our flies without the angler ever suspecting it, inhaling them on a cushion of water so they are never actually in contact with the mouthparts before being ejected (you can observe this from suitable vantage points under certain conditions, when watching others cover a pool). In days of yore and plenty, for some sportsmen this moment of 'the take' was the main objective, and having closed the deal they handed the rod to a gillie to wind up the details; conversely, I once shared a boat on the Ponoi with an Arizonan called Frostie who lolled in his seat while our guide Alex did all the casting, and waited for the rod to be passed to him, ready connected, as if by his company's switchboard operator.

Across our sport as a whole, the spectrum of takes is impressively broad. Off the Pembroke coast I was surprised at the gingerly fashion with which blue sharks would nibble at our (mackerel-last-tipped) streamers, whereas the buzz-saw grinding I heard from the stern proved to be a large specimen masticating our propeller; and trolling along a Cuban reef I saw something the size of a hearth rug – probably a giant grouper – materialise beneath my plug, then strike and sound before I could wrestle the rod butt into its gimballed 'bucket', the lure springing instantly back with its trebles mangled like some flimsy metal music-stand collapsed beneath a clumsy bassoonist. Angling is not always 'the gentle art'.

Benjamin Franklin may have devised the lightning rod, but the lightning thread comes courtesy of Dylan Thomas. I doubt the showman of Swansea had much time for fishing, but his longest poem, the turbulent fifty-four-stanza phantasmagoria 'Ballad of the Long-Legged Bait', focuses on this crucial moment when a fish strikes and then takes away the first of your line, and captures that quintessential thrill as the 'lightning thread' unspools beneath your thumbs. Characteristically convoluted, and fraught with elliptical diction, the narrative seems to describe an angler trolling a mermaid as bait through rough seas and passages of rough sex. In some manuscript notes he is named Samson Jack, presumably after the large megalithic 'long stone' on the Gower Peninsula, rumoured to have the power of answering intimate questions.

The resulting poem is an erotic farrago of desire and loss, flashing with the Creator's thunderous light, awash with amniotic and glandular imagery lewdly referencing the Ancient Mariner, Captain Ahab, the Garden of Eden, Rimbaud, Yeats, Donne, death and salvation. What thunders away from under his thumbs becomes a burning hotline to the universe, and this thunderline (it is rendered as *fils de foudre* in one French version) transforms the fisherman temporarily into a conductor of certain forces of Nature, like one of Franklin's kites tugging at the dark sky in search of 'electric fire'. I have taken this as my leitmotif because in its evocation of the pulsing, fiery, energetic filament that links you to your adversary I feel we have the figurative centre to the experience of fishing with rod and line. To the fish, this convergence represents a fine line between life and probable death; to me, it has become a kind of lifeline, upon which are strung so many episodes, daydreams, disappointments and bright moments down the years, the individual thread spun by the Fates, and also a guiding principle like the mythical 'clue' that you follow to find your way through life's labyrinth. When you are hooked up, nothing else seems to matter a damn, and you are indeed in a state of bliss; but when the connection breaks off, then it feels like Paradise Lost.

At a New York party given by the jazz critic Marshall W. Stearns in 1950, just a few years before he drank himself to death at the age of thirty-nine, Dylan Thomas vouchsafed that his ballad was actually, 'a description of a gigantic fuck', making it virtually unique in the history of halieutical verses.

We'd better return to small Highland rivers, before this excursus makes me lose my own thread. When it comes to invitations, I'm such a sponger that it's surprising I haven't developed a free-diver's nosebleed. Thanks to the generosity of an indulgent hostess, I have for the past twenty-five years cultivated a modest rapport with the Laxford, 4 miles and twenty-odd pools of varied water flowing past Bens Arkle and Stack in that western area of Sutherland where Norse legend records their gods practised modelling mountains when the earth was still pliable and young (its name derives from the Viking for 'salmon fjord', which is a good start in life). I feel all salmon rivers are beautiful simply by virtue of having salmon in them, but by any standards the Laxford is sublime. Whenever I want to summon up the paradigm of a smaller-scale salmon river, this is the place my mind supplies – working my way, in the company of gillies Robert and John, down from a lie beneath the tree on Top pool, via the mini-canyon at the head of Duke's, through the numinous gorge where, if you look up, you might expect to glimpse a hamadryad or some horned god peering down on your efforts, and so towards the sea. Augustus Grimble (who saw almost every Scottish stream) pronounced it in 1913 'an ideal angler's river', and so it remains.

It is in the nature of spate rivers to disappoint the casual visitor. Scotland receives some 900 billion gallons of annual rain, but too often you find the stream drooling over a rickle of slimy stones, glaring at you from its unhappy bed while the odd manic-depressive resident crashes from its lie to disturb the unwelcome silence. You pray for precipitation, mindful of that Yiddish proverb, 'So many prayers, so few noodles.' One Monday morning (in July 2010), I awoke early to the sound of fat raindrops spattling onto the creeper leaves of the lodge and saw the little garden burn tossing its mane around formerly bare boulders. As we drove to the Laxford itself the crags were feathered with white rills – and, lo, as the psalmist wrote, 'The river of God is full of water.' Conditions became perfect as the week progressed, and I finished with an unprecedented eighteen, including one real 'banner day' when I held onto seven fish – something I usually read about happening to others. But on the Saturday, despite all the confidence spawned by my lucky triumphs – and on a day when my friend Algy grassed eight – I could not engage a single salmon in discussion until, just before close of play (and to the delight of those savouring my frustration), wading below the bridge I hooked a grilse which somehow ran around behind me and flung my fly back in the air.

On rivers big or small, it periodically pays to ring the changes and, although I think that size, depth and speed are generally more crucial than pattern, nonetheless I enjoy fussocking around in my boxes for supposed inspiration, and there may well be merit in showing a line-sick salmon something other than the predictable Cascades and Stoats which it has already encountered elsewhere on its progress upstream. There is also a range of minor tactics that can be brought to bear when a conventional fly fished 'flat' has failed to excite interest, in the time-honoured 'across and down' routine, hand on hip, then taking three steps to recast. For instance, 'backing up' is used too rarely; I first learned it on one of those canally stretches of the Thurso when a wind had made the surface as ruffled as a philosopher's brow, and while the footwork can be laborious it seems to draw upwards some fish that have lain doggo when first covered. Variation of retrieve is something I watch carefully in others: a constant figure-of-eight, slow handlining, rod-tip waggling and bouncing, or the in and out pumping movement of your whole rod (sometimes referred to as the Scandi Wank; I call it, Come Again, Vicar?) are all worth having in your repertoire as variations on a theme, along with a sudden acceleration off the dangle. I have hooked two salmon while sweeping into the upstream manoeuvre of a Spey cast.

It also pays to scratch the surface, especially when the water is warm. Profile flies like the Collie and Sunray can be great locators of reluctant fish, and even if they merely swerve at it you will have excitement. Ray Brooks's original Sunray Shadow measured 5 inches long and was winged with colobus monkey hair from an old carpet: on spate rivers I prefer a mini conehead version, throwing it off an 11-foot single-hander as stern and lissom as my prep-school art mistress. These patterns can also be effective dead-drifted over known lies. Deliberately skated flies, as when dibbling a dropper Helmsdale-fashion, go back at least to the time of Trooper Franck ('Dibble but lightly on the surface and you infallibly rise him'); I like a double-hooked Muddler for my dropper, and have twice hooked two salmon at once, but never succeeded in landing both. Hitching has become more popular: originally the Portland Creek Hitch (from the Newfoundland river where flies were first attached with a half-hitch when their gut eyes had perished), this mesmerising method requires absolute focus as your iron rides across the current at an angle, inscribing the surface with an attractive arc. This is sometimes described as 'technical fishing', though I don't see what other types there are – good old chuck and chance it, I suppose.

Some say dry fly-fishing for salmon (using a fully floating fly) began when the Revd Elmer Smith lobbed his cigar butt into the Miramichi and it was engulfed by a fish. The dry fly proper – fished in short, drag-free drifts over observable lies – offers regular sport in North America and Russia, but has never quite caught on over here, despite pilgrimages by such devotees as George La Branche (1925) and the maestro Lee Wulff (1961) who found their techniques rarely worked on our rivers like the Scottish Dee, arguably because of unfavourable temperature conditions. On the Mallart, however, a challenging tributary of the Naver, Derek Knowles pioneered a pattern named the Yellow Dolly, a diminutive hair-skirted fly that consistently excites salmon there. I was shown its use by the Loch Choire stalker Albert Grant, who insisted it be tied on a section of plastic tube from a WD-40 lubricant aerosol, and presented it on a very long leader, inducing drag. When the Crooked pool was crenellated with a nice 'cockle', I moved three fish under his tutelage but was far too quick with my trigger finger: 'That was a bloody disaster,' he said, though eventually I did secure one from Jock's, and Bert gave his grudging approval. Later that year he died on the hill.

If they refuse to come up, you can always go down for them. One further method I resort to, specifically in low water, is the salmon nymph. On the Test they have developed a tungsten-headed monster marabou creation (similar to Trevor Housby's classic Dog Nobbler) which you agitate vertically over deep-lying salmon, and this can work up north as well; I used it with some success on the lower stretches of the Halladale when lots of fish were packed in awaiting rain, though they have a tendency to attack it with their tails. Smaller, less rebarbative nymphs cast upstream over spate river pots can be surprisingly effective, if you have the confidence to deploy little hooks.

Back in July, on the lovely Laxford, we are on the Lower beat and the water is fining down after a spate. Somehow the air seems differently charged this morning, the light has thickened, and it feels as if conditions might all conspire auspiciously – anglers sometimes speak of a 'sixth sense' that things are about to work out, and I have this vague presentiment we will succeed (though such inklings often prove wrong). The current on the Lower Bridge pool is muscled and gleaming as an athlete's thigh. Robert proposes a Collie, and although some fresh fish are showing (we are not far from tidal water here) they will not respond. We try the Silver Stoat and Munro Killer, but no deal. Then I show him a Red Frances, that flamboyant black-eyed prawn pattern that has become an Icelandic staple – fishing a pool with one is sometimes called 'Dicking it', a vulgar pun on thriller writer Dick Francis. By now the little pool is jostling with fish, like a stew pond. I lob the heavy tube upstream and third cast comes a take like a bodybuilder's handshake, and the fight is on. Seldom have I had such a spirited encounter with a Scottish fish – part gavotte, part tug of war – as first it turns for the salt, then attempts to run the rapids at the neck of the pool. Eventually, Robert lifts the net and we measure her at just over 34 inches – a hen weighing a good 18 pounds, and a trophy for this most bountiful of streams. Nowadays I seldom fly-fish without a Frances somewhere within easy reach.

It was on the Laxford that the duke's butler once announced during dinner, 'Your Grace, I have had word that the salmon have begun running the river.' So Bendor Westminster left the table and boarded his yacht: the river referred to was in Norway.

I have never landed a huge salmon from Norway, but it remains the place to go big-fish hunting and has been so since the 'salmon lords' from England began exploring its potential in the Victorian era. In 1848 a remarkable book, *Jones's Guide to Norway*, was published by a jobbing hack named Frederic Tolfrey, who had never set foot in the country but still furnished confident advice for sporting visitors, including their stopover in Hanover: 'Those who love pretty girls, and take delight in saltatory exercise, will find plenty of amusement in these *soirées dansantes*.' Doubtless the milords' manservants ensured dancing pumps were packed in the steamer trunks, plus something for the weekend. The sport could be exceptional: on the Alta (where a 20-pounder is regarded as a grilse) that same duke landed thirty-three fish in one night. He also took a fancy to the Reisa – which I fished without result, though one casting champion was pulled bodily into the river when his line jammed while battling a storlax – and in characteristic style, paramour Coco Chanel in tow, in 1927 decided to construct a lodge way upstream, which he only ever visited for a single day. (I once slept in Coco's bed – alas, several decades too late.)

Impressions of the few Norwegian rivers I have tried – Lakselv, Børselv, Maselv – are a mosaic of cliffscapes worthy of Doré, evening *rund opp* sessions with shot glasses of fierce cognac, bankside coffee from a Lars kettle, oxblood-red farm buildings, reindeer soup with crispbread, chainsaw snoring through the thin walls of various lodges, and the sight – several times, making me flush even now as I recall them – of vast fish wallowing at my heavily coiffed tube flies as they sashayed over a bouldery residence. In 2012 a group of journalists made a whistle-stop tour of Finnmark, and in Lakselv we were housed in what I took to be the grim civic psychiatric facility, though we were assured by our hosts it was a hostel that had been vacated exclusively for our use. Imagine my consternation when I confidently entered the shower block and encountered a student nakedly intent upon her ablutions. There is a local expression when something strange is going on – 'There are owls in the moss' – but the evening was chiefly memorable for a volcanic take later by a 24-pound fish in Kairanen pool that whacked my Templedog and took off like a freshly fucked female fox fleeing a forest fire (as I believe they say in Colorado). The Norwegian version of 'Tight Lines', an angler's phrase for 'good luck', is *Skitt Fiske*, and it deserves wider currency.

*

In 1914 the last passenger pigeon – Martha – died in Cincinnati zoo. Once, flocks of these birds were so dense they took days to pass overhead. Mass migrations (eels, buffalo, butterflies) are becoming a thing of the past – human refugees seem to be the exception. Early American settlers merely had to lower a bucket into the Atlantic to catch cod. Salmon were once so abundant in the Rhine that parr were used for pig food. The last Thames salmon (until modern times) was sold to George IV by a Mr Finmore, for a guinea a pound. The Doctrine of Plenitude no longer holds water.

These days, if you want to see Atlantic salmon runs as they were in the days of plenty, Iceland is the place. It can boast nearly a hundred salmon streams, which is not bad for an island with the same-sized population as Ealing. Some of these systems record a thousand rod-caught fish a season, partly because many Icelanders still like to use the worm: one such extractor is a Reykjavik dentist nicknamed Toti Tönn, who is said to have accounted for some twenty thousand island salmon.

Ultima Thule certainly *is* 'a land of vivid contrasts'. Edgy, other-worldly, haggard, fissured, greener than its gelid name suggests but with a heart of black sand, it is a geological battleground of ice and fire, littered with volcanic scoria, tufa and screelines cindery black or blood red, where the earth's crust is so thin in places that there is unfinished business (the island of Surtsey only arose from the sea in 1963) and the weather is so unpredictable the joke is they just get samples. God's laboratory is also home to some of the least appetising traditional cuisine – seal flipper, pan-fried guillemot, for example – and even the modish modern chefs dish up minke whale tartare or geothermally baked 'thunder bread'. One specialist brewery produces ale infused with smoked whale testicles, which you won't find in the Borough of Ealing.

I first fished here in 1987, on the Laxá í Leirársveit, one of several rivers I have visited in the west. As someone used perhaps to a couple of Scottish fish a week, nobody had quite prepared me for the possible density of salmon numbers and I landed five the first evening, on tiny doubles. But then I endured two blank days, because I had much to learn – particularly about the need for precise casting, often tight beneath the opposite bank, the perils of skylining, and the need to work out your angles of approach. A belted earl upped my game by giving me a large lure named Pat's Pussy, which worked well in the foss pools. Targeting salmon by sight has become a favourite ploy on subsequent trips: in the dramatic Solheimafoss on the Laxá í Dölum (which I fished in 2014) there were echelons of Atlantics visible from the clifftop, and you could cast like a berserker with weighty Snaeldas and love bomb the odd one into snapping at the lure. Also here I was shown how to fish a micropattern off a clear intermediate 'slime line' – little Madelines or Frances trebles twitched rapidly back, a method not often tried in Scotland.

It was truly MacIntosh weather in September 2016 when I stayed on the Laxá í Kjós, nicknamed the 'salmon angler's university' (graduates have included Jack Hemingway and Kevin Costner), but our guide Sibbi, typical of the enterprising young

Icelanders who help you out, took me up to Pool 35 and rigged a two-nymph leader with which I caught a brace of sea trout weighing 7 and 4 pounds, demonstrating again the advantage of a little lateral fishing. For the last week of that season we went on to the fabled Grímsá – 'Queen of the Rivers' – with a modernist lodge designed by angling author Ernie Schwiebert (sometimes unkindly known as the Elephant House). The first pool I fished was named Horrible, after a monstrous creature that was netted here back in the time of the Sagas. Quite often you find topographical spots that are mentioned in these mediaeval prose tales of magic, incest and revenge (saga means 'something said'): on the Haffjarðará, where lava outcrops seem to watch you like orcs, Grettir's pool commemorates where the red-haired outlaw flogged an enemy with a tree branch, and the country's mythology so teems with fishy, shapeshifting lore that in places you feel you're casting deep into the nation's collective psyche.

Doddi the guide bade me hitch a Langa Fancy initially down Gardafljot, and then as dusk approached we switched to a small Sunray in the bulging golden light. Salmon were head-and-tailing along the far bank, and up they came obligingly. It was the fishological hour: a pearl harvest moon appeared in the sky simultaneously with the blood balloon of the sinking sun, and as I strip-struck the fourth rise, there came a Satchmo growl from my Bogdan and a sizeable coloured fish began greyhounding like a marlin, my fly shaking darkly in his scissors. Hoisted briefly for the camera, lit by apricot and nacre, he was beaked like Punchinello, and at 18 pounds proved the best off the river that year. Later, with staff and guides, we celebrated the season's end with Lagavulin and air guitar, and one guest swore he saw the Northern Lights. Something said, indeed.

We were drinking White Grizzly cocktails – a real grapple-the-rails combining vodka with Caucasian champagne – in the rooftop bar of the Hotel Metropole, Moscow, in late spring 1974. It was the last evening of my gap-year travels to the Far East, and James (one of my more grown-up friends, who later became a government minister) had already put up with my protracted food poisoning on the Trans-Siberian Express from Nakhodka; now, as a Russian speaker, he was trying to advise me that the formidably built woman against whom I was propped on the tiny dancefloor was a policewoman, and that her husband had recently entered the room. I recall being sick in several languages into a bank of dirty snow as we slunk back through Red Square.

When I eventually went on my first fishing trip to the Kola Peninsula in 1991, therefore, I had some congealed memories of Soviet life – mineral water that tasted like hookah dregs, toilet tissue apparently sandpaper on one side but waxed on the other, and a ubiquitous smell combining hot plastic, plum jam and a dash of drains. But I was thrilled to be returning because the Kola (which has some sixty-five rivers) had been a military exclusion zone for ages and was just being opened up as a sportsman's ichthyotopia.

Over miles of taiga country, jigsawed flatlands, dwarf

birch and moss banks, the big old chopper, a veteran gunship from the Afghan war, ferried us to a makeshift new camp 40 miles upstream from the White Sea on the Middle Umba. The fishing wasn't much good – in August we were between runs – and the terrain was strenuous, like yomping over ground as spongy as a mad cow's cerebellum, with mozzies large enough to hump a turkey, and no boats for crossing the pools. Tolya, the head guide (whose wife Olga, a comely Muscovite, toiled behind the kitchen hatch), spent much of his time standing proudly on the shore in his new black-market trainers, bellowing, 'Iss big fishes further in, Dayvee,' as one waded perilously along the drop-offs in search of the odd red resident. The prospect of virgin water is as sublime as the promise of love, but in fact this was far from *aqua intacta*; every summer a number of poachers (known as 'brucaneers') brave the hardships of the taiga to fish for the pot, and one could readily discern many of their favoured spots by the scraps of blue insulating tape near the tails of pools. The 'brucs' would arrive surreptitiously, cut a sapling, and tape on some rod rings from their pocket, before lobbing in a bait or spinner. One morning I found an improvised rod with a rudimentary fly attached; I snipped this off and tied on a fully dressed Green Highlander for the chap when he returned to claim his gear. Occasionally a fisheries officer would discharge his rifle randomly into the woods on the far bank: it was A Far Cry from Speyside.

Washing was limited to a bankside *banya* tent, within which one was beaten, Grettir-style, with scalding birch branches, before plunging into the Home pool. Despite being slathered in deet, the aftershave of the Arctic, the accumulated mosquito toxin was so intense that by the third day my visage was swollen like the Elephant Man's, and I had to loosen the strap of my wristwatch by two holes. Stopping for a ciggie break after an hour's walk in chest waders, guide Valera gave me a powerful papirosa and indicated a concrete foundation slab nearby, jutting from the undergrowth: this was the site of a former Gulag logging camp where, he informed me neutrally, the punishment for unruly prisoners was to be smeared in honey and roped to a tree in the sun until you died from insect bites. After that, we stopped complaining.

When the daytime proved too glary, you fished during the 'white nights': even at three in the morning it was light enough to read a newspaper outside (if you had one), and so it was that on the Fourth of July, a date not much celebrated in these parts, my friend Simon and I walked down to Policeman's pool in the early hours of the morning, where the river divides into an isthmus of five finger channels debouching into a lake. It was here that I heard the singing.

I am not really prone to flights of fancy, but there came a wash of high, ethereal female voices dissolved upon the breeze – siren music, the words of which I could not discern, cold and beautiful, distant yet distinct. If there were people there, ordinary eyes such as mine could not see them. It was lovely rather than sinister. The entire air was a harmonic mesh of entrancing sound, like Prospero's cove. Perhaps I was spellbound – certainly I ceased casting and stood in the rustling current for a good while before reeling up and walking back upstream, where Simon waded out of his glide and merely said, 'You heard it, too, then?' and we headed back to the huts in silence.

For centuries, semi-nomadic peoples must have stopped each summer at these liquid crossroads to set nets for their vital harvest. There are archaeological signs of a Kola culture dating back to the sixth millennium BC. Burial objects include fishbones for the afterlife; and granite cairns, often quite elaborate, were erected to mark productive spots to hunt and fish. Here, the split river was conceivably acting like some kind of node – a place where the barrier between two times was less than substantial, and the succession of those once here working their nets and singing their ritual chants must somehow have left behind a psychic resonance, recorded in the rocks, a soundtrack of ages long past. Perhaps J. W. Dunne could have explained it. Maybe there were owls in the moss. I don't know, and I have never again heard anything remotely like it, though years later I described this experience to a Saami woman (in the Outer Hebrides, of all places) and she was not surprised by any of the details. As the canny H. T. Sheringham observed, 'It is just as well to have a good conscience if you are going to do much night-fishing.'

It was on the Fourth of July 2001 that I truly met Mother Russia up close and personal, and we nearly became inextricably wed.

Our journey out to the Varzina had not progressed smoothly, and we were fogbound for a day in Murmansk. A tenebrous Arctic port of unlovely concrete architecture, famed for its smoked halibut, the city is not exactly twinned with anywhere but is said to enjoy a suicide pact with Scunthorpe. In the saloon of the Hotel Poliarnie Zori ('Your Far Away Home' croons the brochure, though it might as well run, 'Ya Gotta Stay Somewhere'), there soon foregathered a drove of colourful female souvenir sellers, eager to provide our gaggle of rods in fleecies with some saltatory exercise and a keepsake or two that might subsequently require medication.

The Varzina is a variegated river with a reputation for big fish. A 20-pounder (or 'dolfinn') earns you a spiffy badge, but there are two grades up: an English angler of integrity and repute (sadly cameraless) once landed a salmon 'closer to eighty pounds than seventy', measured it on the riverine pebbles and later had a mighty plaster cast made up. One afternoon I hooked into a fish at the vee in the tail-out of glorious Hourglass pool, and it simply turned and bolted down the long rapids heading for the Barents Sea ('pining for the fjords' as they say), whereupon my massive guide – sporting a stripy Spetsnaz T-shirt – hoisted me in a bear hug and carried me along the bouldery bank for a quarter of a mile before netting the exhausted fish. The surrounding woods were loud with cuckoos uttering their 'uh-ohs'.

At nine-thirty-five on the fourth, our camp's Mi-2 helicopter (flight RA-23380) took off for a day trip to the nearby Sidorovka river, with two crew and six passengers aboard. I climbed into the co-pilot's seat – we took it in turns to ride shotgun, for the God's-eye view – and duly buckled up.

Blu-tacked to the dashboard was a gleaming feather – 'Is lucky eagle!' explained our pilot, thumb up. Twenty minutes later, translating into landing mode through a canyon, our aircraft was swept into some kind of vicious downdraft and as we hurtled at high speed towards the rocks I remember, terrified, shouting, 'Jesus God' just before the machine hit the rocks nose-first, slid 80 yards on its side in a welter of tearing metal, shearing off rotors and spouting fuel, the rockscape thundering past my ear (which was by then very close to the ground) before coming to rest in an impressive debris field.

I was on the underside. There was a small fire behind my head, which the crewman, despite a broken hip, valiantly crawled to extinguish. Crunching several of my ribs, the pilot stood on me and punched out his side door, which was pointing at the sky. Miraculously there had been no explosion on impact, and everyone made it out. Worst off was my room-mate Paddy (a former RAF officer and father of guide Sean) who sat on a peat bank, putty faced and with thick blood bubbling from the corner of his mouth (he had punctured a lung). 'I think I've had it, old chap,' he said. I passed out. 'Come on, let's clean you up,' said kindly Trevor, washing my face where blood was leaking from my scalp, gashed by a shard of fuselage. (I was lucky to escape with three broken ribs, and a zigzag scar below the hairline – not quite a Harry Potter lightning mark, but close enough for comfort.) We took refuge in the nearby bothy. There were a few breakages, but nobody died. We wrote painkiller dosages and times on the foreheads of the injured, to assist any emergency services who might appear. Gallows humour prevailed. We had a luncheon hamper and some Scotch, and I said to fellow rod Joerg, 'By the looks of things, I won't have to start eating any of you before Thursday.' Paddy gamely remarked that for years the Russians had been longing for him to have a prang, and it had to happen on his first ever fishing holiday. (Months later, I asked him if his old aviation colleagues back in Blighty had been sympathetic about our uncontrolled flight into terrain. 'Not really,' he replied, 'but I think there was a sweepstake on whether or not I'd lost my monocle.')

From the wreckage I salvaged a heat-blistered section of rotor blade for my tackle-room shrine, and there in the soil beside the cataracted windshield was the eagle's feather, also now in my collection, a memento of our Russian Grand Slam.

The emergency beacon went unanswered – safety procedures have improved immeasurably since then – so it was ten hours before we were rescued. Five survivors would be evacuated to Murmansk General Hospital, and eventually the FSB investigated the crash. But for now there was fishing to be done, so three of us cannibalised sections from our various shattered rods, assembled a Sage-Loomis-Hardy Frankenstein model, and headed for the river. 'Careful on those rocks,' I called to Joerg, 'you don't want to slip and have an accident.' The name of the pool was Paradise. *Et in Arcadia ego.*

13

OF GILLIES AND GUIDELINES

'Jamesie was a red Celt. Time had sown sand over
his flaming hair, and taken some of the colour from
the blue of his eyes, but his back was still as flat and
straight as a door, and he walked delicately, like a
cat. He admitted to being eighty ...'

T. C. Kingsmill Moore, *A Man May Fish* (1960)

Some sportsmen prefer to fish solo, but I have almost always
found the presence of an expert companion an additional
pleasure, and one that maximises the chances of getting
hooked up. Nowadays part chaperone, caddie, chauffeur, bird
dog, shaman, *genius loci*, fish whisperer or *valet de chambre*, the
original gillie (there should be no 'h') in the Land of the Gael –
cas-fliuch – was a servant to his Highland chieftain, one of

whose duties was to carry him bodily across fords. I am only a very minor laird without any retainers, so it is not a service available in our neck of the woods.

In semi-feudal days there was a gillie stereotype, now largely disappeared – a doleful, crabbit attendant with an ineradicable demeanour of gloom, cheeks empurpled, fuelled by blether and ancient grudges, often an incompletely reformed poacher, set in his ways and unimpressed by all comers. In recent times a new generation has evolved, influenced by the freelance professional guides of Northern America, with broader skills and experience of places other than the single stretch of water to which the traditional gillie was riveted, offering a complete service that might include custom-tied flies or tuition in various Spey casts. There are high-tech coarse fishing guides, too, harking back to the Thames professionals (of which there were once hundreds), like Wormy Webb of Reading, who kept his bait in the bathtub. This human factor remains central to the entire angling experience, though I sometimes congratulate myself that during one of my most successful day's angling – seventy-eight cutthroats from a remote stretch of Montana's Slough Creek, apparently then some sort of lodge record – there was no guide with me. Had there been one, perhaps I would have done better.

Gillies and guides are a notably heterogenous group, but having fished with many dozens of them I feel there are certain basic desiderata: your hired help needs to be courteous, attentive, observant, minimally sarcastic, intimately acquainted with the water, neither too taciturn nor garrulous ('vaccinated with a gramophone needle', I heard it called on the Mourne), steady in a crisis (feelings can run high), not too thirsty (gillie is by chance also the word for a measure of liquor, cognate with gill), discreet, diplomatic, reasonably encouraging, able to maintain morale when sport is slow, punctual, resourceful, and a tolerant listener – despite years of being exposed to piscatorial tommyrot and horse feathers. An ideal gillie is a paragon who appears to find intriguing your paltry reminiscences; meanwhile his patience with your frequent bawks and fankles and 'wind' knots must be unremitting, and he should not be one of those self-appointed 'characters' who persist in regaling you with anecdotes of how well Sir Victor Spatchcock once did on this very pool ('He was a verra guid fisherr, mind'). As Chekhov wrote, 'The most intolerable people are provincial celebrities.'

If you travel widely to fish, you will inevitably encounter the odd charlatan, dud or shitweasel for whom everything is too much trouble, but I have found them remarkably few. Juiceheads are increasingly rare, though once I did have to wait two days for a Christmas Island guide to sober up (one of his colleagues had the habit of opening his morning bowels uptide of our wading line, which seldom added value). Lechery is another no-no. One lady friend was subject to amorous overtures from a Russian guide up in the Arctic (where the breeding season is, admittedly, short) but she solved the situation with a swift kick in his kukushkas. Indolence used to be common. One youngish Spey gillie would sit on the bankside reading his paper with no intention of coming to one's aid, whereas I require a plausibly amphibious assistant (an indomitable Dee gillie I know once plunged into a pool to rescue an angler who was being swept away, and subsequently married her). Another Spey veteran had developed a special trick of the trade: he would mark down a salmon that looked likely to be a taker, and just before you reached the spot would ask if he could try a few throws with your rod, often promptly hooking up and thus appearing to possess piscatorial superpowers. Insolence is a grey area, though I believe Ben on the Thurso meant no harm when he enquired if I was a Freemason because of my casting action. Just don't expect admiration; and never try to impress a Scotsman. One Glaswegian, upon being shown the Victoria Falls, was asked if he had ever seen anything so remarkable and replied, 'Weel, ah ken a boy in Paisley keeps a peacock wi' a wooden leg.'

Some foreign guides are mavericks – idiosyncratic loners who don't always play well with others, and many regard their job as a form of babysitting (Live Client Out, Live Client Back: Job Done). In Alaska, lodges often arrange a guide-training week, during which the old-timers simulate some antics and mishaps of their more nightmarish clients – falling overboard, snagging logs – and guide skills have to include lifesaving as well as the preparation of shore lunches. Certain Scottish gillies of the old school would be surprised by the level of service offered by their overseas counterparts.

Having neither the expertise nor the temperament for it, I have seldom been much in demand for guiding, but I know it can be challenging work. Apart from resisting the impulse to throttle your patronising and incompetent client (which generates time-wasting paperwork), it can be physically tough, as anyone trying to wrestle a loch boat all day in a high wind, or poling a flats skiff under broiling skies, can attest – the only time I tried it, in the Florida Keys, I toppled off the kicker bridge into some painful mangroves. I have attempted to teach a few people to fish, but there's a difference between showing and teaching, and I certainly have not yet arrived at that stage where I derive as much satisfaction from seeing others catch fish as doing it myself. On the Ponoi Home pool after dinner one evening, backlit by a setting sun, I went in deep to net a friend's large cock salmon and he said, 'When you came back out of the water, you looked like the Guide from Hell' – one of the more nuanced compliments of my angling career.

Your relationship with a guide can be delicate, and there is some etiquette involved – the client needs to have certain qualities as well, if things are going to prove harmonious over any period of time. Politeness makes a good start. I am still amazed at the casual, or high-handed boardroom attitude a few rods have towards their companions, turkeycocking around about their achievements at other expensive destinations. Gillies and their ilk read their 'sports' quite quickly, and know from experience that not every successful business mogul is *necessarily* a successful human being. Nobody likes a wise guy. Modesty is the better policy, and I tend to make it clear that I am there to do my best and will welcome advice, at least in the first instance – and I'd prefer nice clear instructions, not just a gillie's lackadaisical directions to, 'Go up round the corner to the third ash tree, and get your line well over beyond yon stane.' Most guides say they prefer looking after female anglers because they nearly always listen more attentively and are prepared to take instruction. Sometimes it's better not to know what guides really think of you (I learned this as a boy by listening aboard the beaters' wagon to opinions about the 'guns'). Whingers are universally loathed. Don't berate your helpfellow for the weather or failure of insect hatches. Just because you are on holiday, don't expect voluntary overtime. It is often appreciated if you do not stick to hard-core discussions about angling techniques all day long, though personally I try to avoid politics or religion in any conversations, anywhere.

My childhood cynosure was George Murray of Rogart, gone to his 'long home' these forty years but still my measure of others. He had quiet, impeccable manners, eyes like an osprey and one's best interests always at heart. His company added greatly to my time spent on the water, and oversaw a number of my formative fishological moments. Such gillies also furnish a narrative link with the places where you pursue your quarry (Richard Adams guided on the Canadian Matapédia for some eighty years); they are the voices of the stream, and from them you can learn much that cannot conventionally be taught.

In the spring of 1980, I took my bride to India on honeymoon, packing as a precaution a Hardy Smuggler fly-rod as I knew from experience the Himalayan trout season was about to open, and we were heading for Kashmir. We stayed on Lake Nagin in a houseboat run by a suave and brilliantined entrepreneur named Ravi ('Oh my God, the British are like my father and my mother'), and it was indeed a romantic vessel, panelled in budloo pine, with chandeliers and antimacassars. As you sipped your sundowner on deck there came a throng of waterborne hawkers – 'Shantush shawls, only look please, no need to buy'; 'silverware, sahib, boiled in apricot juice for beautiful sheen' – while from the adjoining cook boat came aromas of Irish stew and chicken fricassee, dishes the *khansamah* recalled being favourites with the snipe-shooting Raj officers of yesteryear. (One morning he served us a fragrant kedgeree, and I prayed the fish flesh did not come from the merdiferous carps that greedily glooped around the nightsoil pipes, amid the innocent lotus blossoms.)

By day, I sullenly endured seeing sights with my betrothed – saffron markets, temple shrines, a rickshaw trip to the Shalimar Gardens. Kashmir has long been described as a paradise: 'If there is a heaven on earth,' trilled Emperor Jahangir, 'it's here, it's here, it's here.' I asked Ravi about the fishing. Admiringly, he appraised my gear (there was a nice LRH lightweight reel as well) and professed an abiding love of the pastime. Picnic hampers were loaded into a black Ambassador taxi with whitewall tyres. A small, elderly shikari was engaged to be our guide. Then we were off for two days of marital bliss – first to the Liddar, a stream 7,000 feet up flowing milky with snowmelt (what I call 'ghost water'), where I spent all morning happily hurling a Peacock streamer into a series of honey holes, my ratchet going 'zook', while my stultified spouse (convinced, not for the first time, that I might not be quite sixteen annas to the rupee) dutifully applauded as we tallied up a leash of recently stocked brownies, and then on to Kokernag, which was where the guide repeatedly had to carry me across the swollen currents on his back. Whether or not he, too, regarded me as his mother and father I never could tell, but that chap really earned his baksheesh.

There came an arranged marriage I had not anticipated. In Srinagar, Ravi contrived to paddle us to his 'museum of carpets' – the inevitable cousin's showroom – where my beloved's fancy was taken by an embroidered *namada* bedspread, ideal for our spare room (sucking on my sweet tea, I whispered, 'Though actually, sugarplum, we don't have any spare room, as such'). The barter was agreed: Alnwick's finest, in exchange for a few folds of mass-produced cloth. I still recall Ravi's thin smile of triumph as he graciously accepted my rod and reel, the little pusbucket. Back home, perhaps predictably, our bedspread proved 'all the wrong colours', and was consigned to an attic. Decades of heated exchanges have since churned up in the lava lamp of matrimony. Children, then their children, have arrived. Millions of keystrokes have been typed. But I never possessed another Smuggler, nor have I returned to lovely Jammu and Kashmir, now riven with violence. And no other guide has ever carried me across the stream like a chieftain.

*

I imagine the noun of assemblage for a collection of gillies would be a 'cast', but I can't help thinking of the ones I have met as comprising a rogues' gallery. Some regard themselves as a race apart, and even operate on a hereditary principle, none more so than the boatmen of Tweed. It has never paid to get on the wrong side of these gentlemen; the often irascible Hugh Falkus made the mistake of riling Bert McElrath, doyen of the famed Sprouston beat and himself head of a boatman dynasty, whereupon 'Falk' was unwittingly rowed all morning over a stretch of water that had never in living memory yielded a single salmon. In boats on big rivers you must have teamwork, as I have discovered with the help of such fine companions as Dryburgh's indefatigable George or endlessly resourceful Rod, who years later still seem to recall details of the exact spots where you hooked fish together, the congratulatory handshakes and the sacramental drams.

On Mertoun in the early days, William Scrope was borne bodily to his boat by John Haliburton, before he finally acquired a pair of costly wading brogues from that politically radical cobbler Younger – who indignantly maintained the boatman caught many of the fish, uncredited, and 'stood only as a nameless auxiliary'. During my many seasons at Mertoun we fished with Jim and John (both sons of auxiliary fathers), and their knowledge brought us to many victories, though at the price of inevitable banter. One morning when I was bungling my casts in an upstream wind, John reported, 'Mister David was fishing like a man with no hands.' His assistant, Ray, a moustachioed former policeman, always dapper in his

Norfolk coat and with a touch about him of Archie Rice from *The Entertainer*, was a fund of wise saws, remarking when you hooked a snag, 'That's the stone he lies behind, sah!' Some of the most memorable days of my life were spent with these kindly professionals.

On stillwaters, your guide can similarly make all the difference to an expedition – not just by coaxing the boat over largely unseen lies but also by ensuring all three of you act in concert, orchestrating conversation and establishing the contours of your day. In the Scottish islands I have come to appreciate this most, though it was on Grimersta in Lewis that I was looked after by the only gillie I have met who had never much cared for fishing himself. (Most adore it, and fish during their own holidays: one Highlander I know sometimes even forgets to ask prior permission before visiting his chosen rivers.) Kenny John was the oldest incumbent of the Asylum (as the gillies' accommodation there was called) and one day he somehow lost his false teeth overboard. Later, a pair was discovered on the shore, and he tried them, but shook his head, saying, 'No. These are not mine.' I believe he was related to a retired island keeper who was once hauled before the sheriff's court on charges of having the body parts of an eagle in his freezer, his defence being that he had wanted to eat it. Asked what it tasted like, he replied, 'Well, something between a corncrake and a sparrowhawk, your Honour.'

There is an unfortunate tradition of sappy writing about the gillies of Ireland, though it is to Professor Luce's anonymous companion that we owe the 'fishological' appellation itself, and he records his Conn gillie, Michael, saying with some wisdom, 'Any change now would be a good thing, even a change for the worse.' But to my mind the finest evocation of such men came from the Irish Supreme Court judge Kingsmill Moore, who gave us the affectionate portrait of Jamesie (quoted at the head of this chapter) whom he first met on the Corrib in 1926 – already an old man, renowned as a wizard of the wet fly, 'Jamesie owned the lake, if not indeed the earth.' Sharp-witted, occasionally duplicitous, but nonetheless a loyal sparring-partner, he was once motoring with an unaccustomed outboard, saying, 'Sure, I know every rock there is in the lake,' when his keel was struck violently, and he exclaimed, 'And that's one of them!' After years afloat together, Moore was busy with a murder trial when Jamesie died and so missed the old rogue's funeral.

When it comes to charter-boat skippers at sea, it pays to lay out your money with care. The best operate crews that do everything for you, if that is what you desire – rig the ballyhoo in its bridle, co-ordinate the outriggers, tease up fish, set the hook and hand you the bending rod, with cooled drinks to follow. But in the smaller outlying harbours of the big game and reef-fishing world I have come across the odd shyster eager to take a gullible civilian's cash, and you can identify him because he's the only one still tied up to the dock and he's wearing the lopsided grin of some crazed bubblegum republic dictator. Off Antigua once, in a craft largely held together with duct tape, I was trolled fruitlessly around the drop-off by a tonto Vietnam vet, complete with head bandana, The Doors playing on the speakers and a Gran Corona-sized spliff hanging from his lips. It was clear that it was no longer fish he was searching for. The buccaneering lifestyle can take its toll: one chain-smoking Kenyan skipper spent most of our evening in the club bar coughing up blood, and such afflictions can be a touch unnerving because out there during the day your life is effectively in their hands.

Since what they do for a living is genuinely precarious, saltwater pros seem especially prone to cultivating their own legends – few more so than Gregorio Fuentes, who had been the replacement first mate for Ernest Hemingway on his beloved Cuban-based *Pilar*. Papa would, I guess, have been one tough client to work for, being intolerant, bombastic, temperamental and egocentric, but he seemed to inspire loyalty in others who shared his lust for death (one of his African white hunters told me Hemingway was the bravest man he had ever known). When I met Fuentes in the village of Cojimar, he was already ninety-eight and making a good living from quayside photo opportunities, since he liked to claim (misleadingly) that he was the role model for the original Old Man. Instead of dollar bills he sometimes accepted a free meal at the local *paladar* (La Terrazza), thus acquiring the nickname 'The Old Man and the Seafood'. He smoked incessant cigars, and was said to be 104 when he died.

It was, however, a Venezuelan tarpon guide who went beyond the call of duty during an exploratory visit to Rio

Chico. This was my first introduction to the tidal mangrove back country – a mephitic lagoon thick with plastic jetsam and crawling with red and black ceramic-looking crabs, the overall reek suggesting some gongfermor's leather jerkin. The place was an extensive nursery for baby 'poons, and we were busy all day with our 9-weights and Cucaracha streamers, as Uffalo punted us around the steaming glades. I boated one fish of 20 pounds, which was considered a biggish moo-moo for that parish, so he truncheoned it for his supper. He then invited me to share this feast in his little *bidonville* of corrugated iron shacks. First we were served home-made rum by his wife, who next lumbered in with a tureen of what seemed to be brackish lagoon water clogged with lumps of grey flesh (some still attached to silvery scales), the surface of this ghastly broth gleaming with circlets of fish oil and the whole exhaling a whiff of diesel and something herbaceous that might have been mangrove leaf. *'Sopa di sabalo,'* she beamed proudly. I had tasted nothing quite so loathly since Big Mary's special steamed haddock-skin pudding, back at my boarding school, but politeness compelled me to spoon down several ladlesful abetted by beakers of moonshine *ron* being poured by Uffalo's daughter, a barefoot teenager arrayed only in a long T-shirt and some emerald-green toenail polish. 'Marisol,' gestured her father, and though I expect something was being lost in translation – I have barely combat Spanish at best, and by now was slurring in Desperanto – it seemed she was offering us some musical entertainment, whereupon I was constrained to lurch outside and be elaborately sick into the noisome darkness of the backwater, thus putting an end to that particular idyll.

Probably the most intimate and intricate relationship with a guide involves fishing the saltwater flats, and there is something peculiarly pleasing about that synergy when it works really well. Although inclined to be hard taskmasters, their navigational skills and sheer versatility with the push-pole can make a *gondoliero*'s dexterity look like a doddle, and without the quarry-spotting assistance of a skiff guide many of my visits to Mangrovia would have been horribly fishless.

For the last twenty years, Cuban colossus Pedro has been my bonefish guide in the Gardens of the Queen. In the early days he did indeed resemble a Freddie Mercury on steroids – a strapping, bronzed renegade with a Zapata moustache and a grin like the shifting of a sarcophagus lid. Infamous for his impetuosity, apparent misanthropy and tempestuous temperament he has (unfairly) been dubbed the Caliban of the Jardines, and I doubt he has a charm school diploma hanging on his bunkhouse wall, yet he is agreed even by his fellow guides to be almost preternatural in his ability to locate fish, and even if one's friendship with him is shaded with Stockholm syndrome, he makes a most spirited companion out in those far-flung spaces. It amused him to pick on my boat partner Don Jon (who is fluent in Castilian, and also a handier angler than me) and lambast him for supposed shortcomings, sometimes of a personal nature, whereas I was treated as the teacher's pet. One morning, frustrated by our low strike rate, he scampered furiously down the gunwales of the skiff, wrestled Jonathan's fly rod from his hands and began to cast (quite ineffectively) at some feeding bones. Before he married a university professor, Pedro's avowed ambition was said to have been relieving every girl in his home village of her virginity, and it was perhaps with this urgent task in mind that, returning to the mothership on the last day of one trip, he exhibitionistically turned the skiff at high speed into what he thought was a mangrove channel, only to ground us high up in among the jagged vegetation, showered in astonished crabs and saved from severe impalement only by the branch-breaking position of the leaning bar in the bow.

Some Roman emperors during their triumphal processions used to have a slave in their chariot who would whisper reminders that they were still mere mortals. Many guides in the Florida Keys provide this service, too, only they tend to yell it out loud. When subjected to such vituperation ('Ya don't see the freakin' fish? Those shades *are* polarised, right?'), I usually point out that I can get that sort of criticism at home. Mind you, they do spend their days looking after some pretty gonzoid people. They can be intensely competitive, too, and intolerant of missed chances ('Man, that's not gonna work'), but at least you can rely on decisive, if frenetic, instructions ('OK, Dave: lay it down. Two o'clock, forty feet, one cast. Eat, goddamit'). The life of a self-employed guide is fraught with brinkmanship. At a rooftop bar in Islamorada I was introduced to one veteran who had been hooked by his client's back cast with a 2/0 tarpon streamer right in his *membrum virile*, the barb having been struck fully home on the forward stroke. As they were many miles offshore, the phlegmatic captain unholstered his pliers, pushed the hook point all the way through, snipped it off, reversed it and performed an extraction which must have brought tears to his eyes behind those mirrored Costa Del Mars. After that, his fellow guides called him 'sir'.

Maritime appearances can be particularly deceptive, and on a largely non-angling holiday in Eleuthera I managed to engage the services of a local lobster fisherman named Wardell, who claimed to know a good spot for bonefish. In his rolled-up trouser bottoms, lime-green *guayabera* and pollarded straw hat he looked an unpromising figure as we stepped into his rowboat, which he languidly punted round his home bay using a bare branch as his makeshift push-pole (no Loomis or Stiffy products there). In the funky, yellowish water opposite his bungalow we saw three bones all morning, but the smallest of the trio eventually took my Pearl Charlie and when Wardell weighed my fish that evening before making a ceviche of it for his supper it was 10½ pounds – my first double-digit bone, and the most low-tech flats experience of my life. Bahamian guides are not always so mellow: on Andros I was cursed with the caustic Tommy, a part-time preacher full of hellfire rhetoric who excoriated my every mistake ('That was the worst cast I ever seen'). Finally he managed to ground us on a sand-bank during a protracted electrical storm, where we lay prostrate in the wooden skiff's futtocks while he recited charms against lightning. Over the water in the Turks and Caicos I have been fishing for years with the colourful Captain Arthur Deane, dreadlocked and caparisoned like an unemployed pirate, who serenades you with shanties of the saltier type and once exclaimed ecstatically to me, 'You know, I think God must be a bonefisherman.' Amen to that.

My most discreet Bahamian guide accompanied me very early one morning at a camp mostly frequented by divers of the proselytising, touchy-feely, Nemo-kissing variety. YuYu was a native potcake dog who prowled the little harbour stalking the tame, cruising bones, lifting each paw gingerly like some setter closing in through the heather. It was said he had once been lacerated by a stingray tail. He put me on to several fish I would not otherwise have seen, until some scuba enthusiasts en route to the breakfast bar upbraided me for molesting the innocent wildlife, and I had to desist. YuYu was the only guide to whom I have slipped a slice of ham as his pourboire.

Given their traditions of cannibalism, infanticide and human sacrifice (to say nothing of endemic afflictions such as mosquito-borne elephantiasis), the South Sea Islanders must have been a little puzzled by centuries of visitors regularly describing their home as versions of Paradise. 'I was transported into the Garden of Eden,' wrote de Bougainville – and, with its chaplets of frangipani, aphrodisiac perfumes, grass skirts, free love, tattoo, taboo, ukuleles and coconuts (their Tree of Life), it certainly must have seemed idyllic compared with, say, the Falklands. When Cook arrived in Tahiti in 1769 to observe, aptly, the transit of Venus, his sailors procured the insular 'sweets of love' in exchange for the abovementioned ship's nails. Soon came the missionaries, firearms, booze and venereal diseases, and a succession of restless souls, from Melville to a lovestruck Rupert Brooke ('All are one in Paradise'). Gauguin, a former Parisian stockbroker, enjoyed catching tuna, and was told that if a fish was hooked in the lower jaw it meant your woman (*vahine*) was cheating on you; he claimed to have found bliss with a teenager, and eventually perished from syphilis. These days, French Polynesia covers an area the size of Western Europe, with a population roughly that of Nice (though the French are not exactly popular – maybe something to do with the 193 nuclear tests they have conducted in this region).

On an island-hopping press trip to Tahiti in 2015, Helen and I were flying from Papeete to the atoll of Tetiaroa, when the plane's music system began playing Wreckless Eric's 1978 hit 'Whole Wide World', which describes the anxious lyricist's restless global search for the girl of his dreams, who is probably living in Tahiti. I had a feeling in my water that I might be heading for a similar Convergence of the Twain, only with an ideal fish.

This discrete collection of islands (*motus*) was once gifted by Tahitian royalty to a Canadian dentist in lieu of his fees, and was later acquired by Marlon Brando following his work on the remake of *Mutiny on the Bounty* in the early 1960s. Polynesians have notoriously rotten teeth; French legionnaires used to treat their Tahitienne paramours to a set of dentures, but would take the gift back to Europe when they departed, thus diminishing the girls' prospects of future suitors. Some five thousand sets of false teeth were commissioned for the movie's extras, to fabricate those paradise smiles, and there was additionally a fair bit of Original Sin on set – several players had to be treated for what was dubbed 'MGM flu'. Brando fell for his future leading lady Tarita (later also his third wife) when she was washing dishes in a local eatery, and for a while Tetiaroa became their private hideaway. When we arrived, it had recently been converted into a beyond-luxe resort, so exclusive that even the tide had difficulty getting in. The chef hailed from Le Grand Véfour, but you could just slouch barefoot in Dirty Bob's beach bar and guzzle down Love Fizz cocktails if you preferred. There was also wondrously protected flats fishing to be had. This was going to be one tough journalistic assignment.

We were already attuned to 'coconut time' – the islands' ethos is that 'the ripe nut will fall when it's ready' – following a stay in Rangiroa, site of the world's third largest atoll. Here a boat driver named Ugo had taken me to some parts of the reef even he had never visited before, seeking out pods of eager bonefish (*Albula glossodonta*, known here as *io'io*, which would be ideal as a cherished numberplate), though nearly all those we did hook were cut off by cruising blacktip sharks. So, our first morning at The Brando I sit on my cooler box on the flour-white sand of Mermaid beach awaiting my guide and fully expect nothing much to happen for a while as I listen to the distant surf trundling like a goods train and watch some

289

nice bonedogs sidle along the shoreline of this protected, no-fishing bay. But, smack on eight-thirty, a launch rasps around the headland and here is my guide for the next three days, Teihotu Brando, son of Marlon and Tarita. As he loads my gear I see a single bluefin trevally approach: 'Go ahead, let's break the law!' laughs my new companion, and so I make it an offer it could not refuse and we have one fish under our belts before we have even properly set forth. It was squarely hooked in the upper jaw – can we please have that entered in the minutes?

During our first session we spot maybe just a dozen fish, before pulling into the lunchtime shade of Billionaire's Bay (so christened by Leonardo DiCaprio, when similarly in search of *io'io*) – but none of these specimens would have been less than 7 or 8 pounds, and the one I boated must have been pushing 10, a real Moby Bone. Later, my Mantis Shrimp is taken by an enormous singleton – certainly 12 pounds, and the lady of my dreams, my ultimate convergence. 'Maybe one of the biggest I've ever seen,' says Teihotu approvingly, as he nurses her back into the tide. I don't know if Mr Eric ever did track down his fantasy girl, but I feel my own flats destiny has been fulfilled. Over the remaining days I am guided onto golden and striped trevallies and also several giant trevallies, including one weighing 45 pounds that marmalises my Bat Boy (a lure with the profile of a felled conifer) while I'm wading near Rumble Rocks. It certainly seemed like prelapsarian sport.

At first, I did not ask my enigmatic guide anything about his family past. I knew Teihotu had himself been involved in dreadfully public Hollywood complexities, and had returned here to reshape his life and, having had some experience of intrusive personal questions myself, I did not want to spoil our new and pleasant routines. Eventually, he volunteered that he and his father (with whom he was to have a famously embattled relationship) had enjoyed fishing here during his childhood, when the untouched shoals of reef species were truly bountiful. The actor had shown him the way to fillet out puffer-fish livers ('better than foie gras') and the method of wrapping trevally when night fishing so the moonlight would not spoil its flesh. On that last evening outside our private villa with my *vahine* and Love Fizz, as the palm fronds gossiped beneath a sky trembling with stars, a chap could be forgiven for thinking he was practically in heaven.

'These are supposed to be earthly paradises, these South Sea isles,' wrote an especially dyspeptic D. H. Lawrence in a 1922 letter. 'You can have 'em.' Thanks – I think I will.

14

Below the Salt

'One does not know what is under the surface.
There may be something or there may be nothing.
He tries, and the rush of something startles
every nerve.'

W. C. Prime, *I Go A-Fishing* (1873)

For decades I was fishing in black and white, but since discovering how to use a fly in the salt my piscatorial world has since been lit in radiant technicolour.

On 15 February 1990, Bahamian guide David Pinder Jr poled his Dolphin skiff around a mangrove carousel at Deep Water Cay and pointed to an advancing pod of bonefish. I fired off my Pink Puff, came tight on my first ever boney, and as it ran out I realised that the guide had hooked another on his back-up bait rod, handed it to Helen, and our lines had become woefully entangled (not a good thing in any marriage). As she was reeling hers safely in, a lemon shark promptly intercepted my fish, and all I had to show for that first cast at what was to prove my favourite species was a head wearing a look of guillotined astonishment.

Albula vulpes, the whitish fox of the flats, is a lustrous fish who goes by many affectionate names around the subtropics – Houdini fins, His Imperial Shyness, banana fish, John Marrigle, Uptown sucker. An aquadynamic speedster with mirrored flanks – reflective chrome, occasionally with faint olive barring – that camouflage it so perfectly that sometimes all you see is a shadow on the substrate, the boney (its vernacular name stemming from the complex skeletal structure) is so nervy that shoals scatter like smoke when alarmed. They feed by zigzagging across the sparkling shallows, hoovering up shrimp, crablets and fry, leaving behind telltale bluish volcano marks in the sand, especially at the happy hour either side of low tide. Smaller specimens (the finnock of the flats) frequently school up, and scuttle around the sea floor in pursuit of prey, like churchyard pigeons pecking at rice after a wedding. Big specimens tend to be senatorial loners. When they tilt up and 'tail' for food in calm, skinny water this is considered the apogee of small game fishing. 'Strip, strip – stop!' commands your guide, as the phantom of the flats streaks after your fly, and inhales it with a slight shiver of the tail, 'Bone on, man!'

Some things appear to be Platonic ideals, unimprovable – Meerlust Chardonnay, lychees, Garbo, 1970s Gucci loafers, an

Aston Martin DB6 Sports Saloon – and for me the bonefish has it all. He is a creature of transplendence, to borrow O. Henry's wonderful word. I have chased him now in eleven different countries, and of all fish that swim he still makes me the most fin-feverish. Some folk like to wade for him but I prefer the vantage point of a boat. Up in the prow, riding the bonefish express, you are on high alert as you scan the terrain, searching for innuendos – a flash, a puff of silt, the shimmer of 'nervous water' as fish cruise just below the surface. The ground may be porcelain bright, or else a mass of turtle grass punctuated by orange starfish or purple sea fans, so you seem to peer at a Jackson Pollock through ploughed crystal. Rustling over the vegetation, the hull of your skiff makes the long sigh of an exasperated lover. The horizon dances in the heat. Bone fever is cranked up by the furious sunshine, and this is high-speed stuff – there's no musing around on chalkstream benches – so when Señor Macabe scurries into view you must be prepared to fire off your fly perhaps 80 feet at short notice, never once taking your eye off the target. Strip-strike – none of your loch-style lifting – and clear your line from the deck because, even though he's the size of a Scottish sea trout, your quarry is heading off into the next parish and your reel is whizzing like a dentist's drill (in fact, dentally trained Zane Grey once measured a bonedog run at 550 feet). After several valiant sprints – on the 'pancake flats' of Venezuela's Los Roques I have even seen him jump – your gasping prize will come to hand glowing in his own corona of sunlight, his fins showing turquoise tints, wide-eyed and reproachful, but his spirit unbroken. Now, is that beautiful, or what?

In the salt, you need to treat your gear like a blood relation. Veteran 'swoffers' maintain big fish are actually caught the night before, as that is when you should scrupulously clean your outfit, rig new leaders, lube your reel and prepare for the morrow before embarking on that series of life-sized Goombay Smashes. In a skiff, I like to have several rods at the ready, since species variety is often one of the main attractions. The marine tidal flat is frequently a mixed neighbourhood, and apart from the Holy Trinity (bone, permit and tarpon, which, taken together in a single day, make an Inshore Grand Slam, a kind of maritime MacNab) you might have fun with snappers skulking in the mangroves, or a mob of marauding jacks that compete in a Gadarene scramble for your surface popper. Seldom is it humdrum, and you're never quite sure what you are going to see next (which is what makes so much fishing a delight). Unless you are a purist, it is rare to skunk when swoffing. If you have a lifespan of maybe a thousand months, why go someplace where you'll hook just one fish a week? On Ambergris Key in Belize, when the tides were wrong, we once tried hunting down wild pigs with machetes.

I have had a couple of blacktips on the fly, but in practice the proximity of sharks to a flats skiff is about as welcome as a floater in your Pimm's jug. After dark, slounging on the dock beneath a woozy moon with a mighty boat rod, a bottle of Jack and a Montecristo can sometimes make an agreeable diversion, but in daylight I am never happy wading among even the supposedly innocuous nurse variety, and have seen a hooked or recently released bonefish being guzzled too often to appreciate the presence of the Man in the Grey Suit.

There is one dodgy predator I do actively seek out, though. The barracuda is a member of the sea pike family and looks like a physics teacher who has repeatedly been passed over for promotion, with pockmarked cheeks and bad breath. Small ones linger in 'batteries' and can be mistaken for bonedogs until they disobligingly nip off your fly. Bigger individuals lurk patiently like cop cars in wait for an offending motorist, but when they sense their prey the speed with which they strike from a 'standing start' seems almost preternatural. The *bécune* or *gallongo* is the razor boy of Mangrovia and has a long history of GBH against humans; its IQ is around the same as the water temperature (say, 74°F) and it will happily attack skittering or fast-moving baits – many have died with a silver spoon in their mouths. When on the flats I tend to be a winch monkey and fling them a sinuous green surgeon's tube lure that mimics the needlefish, though you can get them to strike a long, braided fly pattern if you strip it in rapidly enough, hand over hand, frenzied-milkmaid fashion. Since he can measure easily a yard long and has a tendency to spring determinedly out of the water in pursuit of an escaping lure, sweep the rod to one side as he nears the boat if you don't want Barry the Bad 'Cuda literally in your face.

Such bonus species frequently save my day. Staying at the highly civilised Delphi Lodge on Abaco in the Bahamas, we were returning through the edgelands to the dock after a tough session on the Marls in high winds, which had churned the bonefish flats milky so that visibility was poor. My Hebridean sparring partner Fergus had explored with his fly rod a little alleyway through the mangroves, and I decided to try my Yo-Zuri popper, which I promptly lobbed into some branches; the next throw landed smack in the centre of the clearing, but my fixed-spool developed an hellacious over-run (I like the French term *une perruque*), during the frenzied unweaving of which a long shadow materialised beneath my innocently bobbing white and red lure. The minute I was able to get it chugging again, there came a series of sabre swipes as Barry annexed the plug, tore around his holt, then fled for the open sea, stripping braid from the reel in violent spurts, like a boar pissing. Eventually I was able to put the heat on him and he was bludgeoned on the head for the barbecue – though personally I never trust them on the plate, as some 'cuda carry ciguatera toxin. He weighed 32 pounds, the then Club record, and a personal best.

Thought to be the place where, sixty-five million years ago, the asteroid landed that exterminated the dinosaurs, the Yucatan peninsula on Mexico's Atlantic coast contains the vast Sian Ka'an biosphere reserve (in Mayan, it means 'birthplace of the sky'), and on four occasions I have visited lodges there, the first being in 1994 when I stayed at Boca Paila, one of the longer-established camps and then advertising itself as 'Grand Slam Central'. Our guide Victor greeted me, a leaky bandage round one forefinger where the tip had been severed by a saltwater croc. A low-slung, laconic Mayan, Victor was asked if he had ever been to the States and replied that he had not even been to Mexico (he belonged to his ancestral heartlands, with their labyrinthine back-country channels and limestone creeks). He did eventually succeed in getting me an elusive *palometa*, as the modestly proportioned permit thereabouts are called, though I suspect some of them are hybrids. The bonefish were not especially selective and were in the smallish scampering *pez ratón* 'rats and mice' category, but it was my first exposure to tarpon that proved a fiasco – over the course of three Mexican trips I lost twenty-two silver princelings, without boating one.

Named for his huge, soulless eye, *Megalops atlanticus* is a dinosaur fish with a rudimentary lung, that gulps air and, particularly around dawn, rolls languidly in lagoons and along the ocean reef. The large specimens (which grow to well over 100 pounds) are generally migrants, and to me qualify as a big game species, though I have never tangled with one long enough to hoist it for the camera, despite having hooked two off Islamorada (known to connoisseurs as Poonville). Smaller relatives lay up as residents, lurking around the mangrove forests. They are curious, exhibitionist omnivores that can tolerate a range of salinity and are widely distributed. In Hispanic regions he is *Señor sabalo*, in Texas 'savanilla' (an unappetising ice-cream flavour), and in Louisiana *grand écaille* for the vast scales, once used as poker chips. A dementoid, harengiform assassin with a mouth lined in breeze block, the tarpon feeds by opening the lid of his head and revealing a maw like a toppled wheelie bin. If you do manage to find a purchase for your hook (some folk stamp on the deck so the fish turns away and you set the iron in his scissors), then he performs an aerial lambada as if on pimp dust, his gill plates making a castanet rattle and showing the blood-red, feathery slashes beneath.

Translating from the bonefish flats to tarpon grounds is like swapping a violin recital for a cage fight. You need a 10- to 12-weight single-handed rod (I like models with a foregrip for extra leverage), and even then a serious unit will give you a prolonged workout. A high proportion of fish 'jumped' will throw the hook, even if you bow to the leap in time-honoured fashion, and you get used to the sight of them mid-vault signing the air with your suddenly slackened line. When blind casting for them in tannin-stained water, you get scant warning before a vicious wrench, and sight-fishing involves a curious sense of dread (do you really want to get involved?) as you see one homing in towards the pulse of your streamer: 'Here comes Johnny!'

At Casa Blanca Lodge on Ascension Bay, my Scottish military friend Lefty and I had some brisk bonefish sessions with a guide (Tomás) who was such a purist that he disapproved of casting at schools of fish or 'muds', yet we managed thirty between us on a couple of days in the popular Tres Marias area. I still had not held onto a tarpon, though, and Tomás took us in a little metal john boat into a lagoon where, to preserve the long-term co-operation of the resident *Megalops*, you were limited to two strikes per person, then back out. Lefty neatly boated a brace in the 20-pound range, and I fluffed my first offer by performing a useless trout strike ('No *trucha*, Dayvee,' moaned Tomás). One final fling, and my home-tied Naomi Campbell fly was taken, I tugged and pumped and side-struck enough to cross that critter's eyes and with a Tarzan yodel of triumph I saw the guide stoop and haul inboard my first ever 'poon, a 35-pounder. There is a shot of me toting it, my sunscreen streaming in the late afternoon swelter, with a bemused look like a slightly singed Buster Keaton. And so I completed the Slowest Slam in the Yucatan. When we turned for home, the sky became a great rampart of coral clouds as the sun began to fall into the sea, the celebratory Corona ale from the cooler was a miracle of icebroth in my mouth, and all was glory.

Fifteen years later, at Playa Blanca (the sister operation at Espiritu Santo), I expanded my tally of flats species with a 'geeker' Slam when, after taking a jack and a 'cuda, my plug was stooped on by a black fish hawk, which I let the guide deal with once it was in the boat. Thanks to a regular visitor there, Rick, a telecom tsar from Oregon, I was also taken, blindfolded, to a secret honey hole deep within the Santa Rosa lagoon known only to two guides, where a canvas-bottomed canoe is stashed and a few times each year a couple of privileged guests are allowed to fish this sanctuary for *robalo*, the common snook.

Resembling a platinum-sided zander, with a protruding Habsburg lower jaw, razor-sharp gill plates, and a dark lateral stripe which gives them the nickname 'linesiders', snook are leery, perciform fish that frequent mangrove bowers, but in Florida can also be taken along the beaches and beneath dock lights up the canals. *Centropomus undecimalis* is hermaphroditic, the males turning to females after maturation, and unusually big ones can grow to 50 pounds. Crouched in the front of our unnervingly small craft, I was packing a 10-weight fly rod equipped with a surface Gurgler (Señor Robalo offers enthusiastic top-water action, but you must not spook the snook). In the shallow, greenish water I saw several industrial-sized specimens, and missed one through sheer overexcitement when it struck; my next hookset almost took guide Fernando's eye out with the rod butt, and I needed every bit of the 80-pound tippet to horse this opponent away from the nearby tangled 'structure', but this was my lucky day and we soon released a snook which was certainly 20 pounds, a rare trophy. I still don't know where we had been, but I felt so happy that night as I bimbled back to my cabana beneath starlight as strong and clear as tequila.

In some places – South Andros, the Roquêno pancake flats – the bonedogs are picky, but in the Florida Keys they all seem to have doctorates in piscatology, as I learned in November '93, when I was the first British angler to be invited to participate in the annual Redbone Celebrity Tournament in Islamorada. I was Fishing Correspondent on the *Daily Telegraph* at the time (only a very modest claim to fame) and I thought it might make an unusual story. I don't know what any readers made of it, but I sure did learn a lot about America's hardball angling culture.

During an eventful practice day out near Flamingo Key, I was knocked to the deck by a flying cormorant, an unhooked tarpon leapt into the skiff and I smashed my new Loomis rod. Over a sundowner at Papa Joe's bar I met flats luminary Stu Apte, so was already feeling out of my league when I attended the inaugural buffet at Lorelei, where seventy-seven teams of three convened in a cliquey display of pastel clothing and glitzy saltwater jewellery. Many of the men sported racoon bands (pale strips of skin below where they have been wearing dark glasses out fishing three hundred days a year) and included stars of the football, baseball and television world. I was introduced to my guide Rick Murphy (later to have his own TV series) and fellow angler D. Scott Deal, a well-known boat manufacturer, and hotshot local sportsman; 'Why, hel-lo,' he said sardonically. 'Of course I recognise you from all your movies.' I guess he had been hoping to be paired with model Lauren Hutton, or that guy from *Hill Street Blues.*

It was a two-day competition in aid of cystic fibrosis, and even by flats standards this was an extraordinarily intense experience because the pressure to hook fish became almost toxic. To qualify, both members of a team had to catch either a red or a bone, and from the start I struggled as we went in search of redfish. Also known as channel bass or red drum (less charitably, 'bonefish without brains'), the red is rather an oafish-looking fish but fights *con fuoco* and is delicious blackened and seared (though the tourney fish were released). You either engage in 'all metal combat', presenting a golden spoon right on their kisser as they push water in the shallows, or, for more points, cast them a fly. Scott clocked up five releases, but my efforts simply would not come together and I began casting like an amateur escapologist, lining fish or pulling my fly out of their jaws. Up on the board we had a total of 75 points that evening, but another guy had amassed 1,750. 'We're road pie,' said Scott. 'He just drove right over us.'

Concentrating on bonefish next day, I was perpetually too slow off the mark with my baitcast shrimp ('Just throw, chief,' urged Rick), so I decided to let Scott take the shots, and he boated three before wading and landing a 10-pounder on the fly. Four bones in one morning is an 'ossum' tally among Keys fish, and my partner would be in with a trophy cup chance – but only if I landed one qualifying fish myself. I felt desperate – never a helpful sensation when you have rod in hand. All lines had to be out by three p.m. and at 2.55 I finally boated a 2-foot-long puppy drum that, in Scott's incredulous phrase, simply 'ate the dogfood outta that fly'. God in heaven, the relief. At the awards dinner he became general division champion, and Rick took top guide, while I resolved never again to try competitive swoffing. On that last, desperate afternoon, as we were drifting a shoreline, I recall a leisure angler sitting nearby on his beer cooler yell to his buddy, 'Hey, ain't that Scott Deal in the boat? But who the hell is the *celebrity*?' There's no doubt my place remains below the salt.

Just above the equator and 1,200 miles south of the fiftieth state is an island so remote it makes its own weather. Discovered by Micronesian people who navigated by following the migration of plovers, or dipping their testicles into the ocean to detect changes in temperature, Christmas Island was so named by Captain Cook when he anchored off the atoll on Christmas Eve 1777 and sent ashore his young officer William Bligh to kill turtles. (These days, the native Gilbertese, having no 's' in their alphabet, call it Kiritami, and their own nation Kiribati.) Father Rougier, a renegade priest, founded the settlement, Paris, a century ago, and there are other small townships such as Poland and Banana, the latter christened when a single example of that fruit successfully grew there. As you leave its limits a sign proclaims FAREWELL BANANA. POPULATION 666, a diabolical number to be advertising. Christmas is not a place of conventional beauty, covered as it is in scuzzy saltbush, infested with land crabs you have to dodge so they don't pop your truck tyres, home to myriad nesting terns, noddies and boobies. It's one of the biggest tureens of fish in the Pacific, but true hospitality is shown by opening a can of imported Hawaiian tuna. Copra, salt and milkfish are the only industries, but since 1996 the neighbouring microstate of Tuvalu has received a considerable income stream from leasing its telephone codes to foreign companies running 'adult' chatline services. When I visited in '94, there were no televisions on the island, though at the weekly luau one guide did sport a Def Leppard T-shirt.

The highest point of Christmas's vast volcanic pipe bowl is just 12 feet above sea level, putting it in the firing line for global warming, though it has been notoriously in the firing line before: as part of Operation Grapple, in April 1958 a thermonuclear weapon was test-detonated here equivalent to three million tons of TNT. (Hiroshima's 'Little Boy' was a mere fifteen thousand tons, by comparison. I have lived much of my life in mathematical darkness, but even I can appreciate that difference.) Islanders were advised to turn their backs to the explosion, though the eyes of a few babies were apparently melted. One zone of the atoll is now called Fry Site, the creepiest name I know for any fishing spot. There is a pithy Polynesian saying: 'The palm tree grows. The coral spreads. But man shall vanish.' As the oceans rise, I'm not sure I'd put money on the trees or coral, either.

Our party of four Brits stayed at the island's then only hotel, Captain Cook (architects: The Three Little Pigs), along with a motley crew of Australian and Japanese anglers and twitchers, plus a single gentleman who was there for the franking of stamps (thus it was I later had an article appear in the *Journal of the Kiribati and Tuvalu Philatelic Society*.) With its year-round season and relatively light angling pressure – no prop scars or push-pole marks here – Christmas was reckoned to be the best venue for bonefish in the world: you might expect to catch in one day as many as a week's worth elsewhere. Much depends on coinciding with the right tides, however: neaps with a high around noon will maximise your chances, and it's advisable to avoid full moons, as fish often concentrate on feeding at night, though I'm told you can sometimes cast for them tailing in the moonlight. You wade all day, and on some of the flats such as Nine Mile or Go Like Hell we frequently had four fish on at once; most were smallish, but earlier that year a record weighing 23 pounds had been taken. I saw one real bruiser on Orvis flat, at first supposing it to be a shark: 'Bon,' guide Tyrone assured me. 'Big bon.' I threw it a blind Rootbeer Charlie, it truffled in, ate and ran for the channel edge, just as an overrun blossomed from my new Loop reel, and he popped the tippet. That would have made me the week's High Hook, for sure, easily *victor ludorum*. Fishing can be such sweet sorrow.

There were also six types of trevally, including the legendary giant (or black), often hunted by meat fishermen armed with bomb-proof spinning rigs. The 'geet' is a double-hard marine bastard which can top 100 pounds and even in the smaller sizes you need a Kevlar glove to handle him. I caught my first *ulua* from a catamaran outside the reef – a mere puppy of 20 pounds, but I was astounded by its pitbull power despite my heavy stand-up rod. Its striped and bluefin cousins would be seen patrolling the flats, and one afternoon I was playing a bone when guide English spotted a bluefin approaching and swapped me my heavier fly rod. I threw the trevally a hot one and thus had my own personal interspecies double-header. Off the beach at dusk I even tried lobbing a gigantic plug off my Ugli-Stik into the surf. When my multiplier jammed on an unseen fish, I was actually dragged down the sand and would perhaps have fulfilled my childhood nightmare of becoming food for ravenous crabs had not my assailant cut off the lightning thread on some coral.

When the once-weekly Air Nauru flight came to collect us a day late, our party was rerouted a cool 2,000 miles in the wrong direction across the International Date Line via the Marshall Islands, and then Hawaii, so I ate four in-flight breakfasts before we arrived back in England. I was wearing a souvenir T-shirt that depicted various types of exotic bird droppings, and one stewardess said to me, 'Honey, by the time you reach home you're gonna feel just like that shirt.'

If Hawaii has regularly been described as a kind of paradise (although the heroic Captain Cook probably never thought so, being mortally stabbed in the neck there, and his body baked to preserve his venerable bones), the same is true of the Seychelles. Gordon of Khartoum was just one pundit who was convinced these islands were the original Garden of Eden – not least because uniquely on the islands of Praslin and Curieuse there grow coco de mer palms, the female variety of which produces a supposedly aphrodisiac double nut – the world's largest – said to resemble a woman's nether regions (the male flower sprouts a vaguely phallic catkin). If this were the Forbidden Fruit, Eve would have required considerable biceps to heft it. Once a notorious pirate haven for the likes of Olivier 'The Buzzard' Levasseur, who handed over a cryptographic treasure map on the scaffold and was an inspiration to R. L. Stevenson, the Seychelles remained an unruly Crown colony until independence in 1976. In October 1957, my father was sent there as undersecretary of state at the Colonial Office to assess the area's tourist potential, and advise on whether or not to establish an air link. 'Many of the Colony's leading personalities are bastards,' he reported, alluding to the 60 per cent illegitimacy rate, and he reckoned that the place might indeed be Eden, 'except for the absence of any serpent'. (In fact, there is one rare and venomous yellow sea snake.) These days, among other attractions, it is a fantasy destination for anglers in search of the exotic, if not quite the forbidden – though the pass key to Paradise can prove costly.

To reach Cosmoledo, our Beechcraft 1900 had first to land on Assumption (once they'd cleared the giant tortoises off the airstrip), then we boarded the *Ocean Explorer* – a converted German minesweeper later fated to be hijacked by Somali pirates – and set sail for this rarely visited and uninhabited atoll 600 miles south of Mahé. Our party included a Dutch physio, two Japanese journalists, a Corsican casino owner and a detective inspector from the Met.

A dozen miles across, and subject to dramatic tidal fluxes, Cosmo is a true venue for *pêche extrême* (it was here we hooked that marlin off the inflatable). Back in 2008 it was still largely an unknown quantity, and the blue-water sport had scarcely been explored, though one thing was for sure – the lagoon was home to a formidable population of geets, and we set about targeting these psychopathic dragbusters in a number of ways. As we walked the outer shoreline of the reef, they could be teased into range by a hookless plug (known as the GT ice cream) being thrown far out to sea by Olivier – our *soigné* French host, not the corsair – which allowed you to hammer your Bushpig fly into the headwind as the fish raced in towards your feet, though the surf was so vigorous I was repeatedly knocked over. Twice we managed triple-headers, but these were mostly 10-pound babies, as opposed to the really big ocean-going marauders, which have been known to yank a man into the air when hooked, or to attack your wading boots (tangle with one of these, and you're really putting yourself on the line). Some mornings on high water inside the atoll, teams of three drifted in a Zodiac RIB, similarly teasing fish (including the odd grouper) close to the little boat. At one point, an approaching geet abruptly swerved and accelerated away as a sizeable hammerhead sashayed into view, which does rather jump you mentally out of gear if you're perched on a rubber raft filled with air.

Around the mothership at night a congregation of trevally lay doggo beneath the floodlights, occasionally crashing to the surface if an unfortunate flying fish bounced into their zone. Once a few had succumbed, the others soon learned to avoid all artificial flies, but I did catch one by cheating: as the kitchen boy emptied his slop-bucket – the contents of which always greatly excited the fish – I lobbed in my streamer tipped with a fried chicken wing, and had a nice hook-up (another chap did the same with an old tea-bag, presumably GT Tips). After a couple of days, the fly-only ethos seemed to relax during our evening ritual of duty-free grog, and Corsican Paul somehow produced a pail of small livebait which we proceeded to freeline below the hull, though even with 130-pound nylon we were almost invariably broken off before we could ever see the fish – '*un camion*', Paul would laugh. 'Here, you need two hundred pound!'

By day we walked the drainage channels and cascade pools

just before low tide and threw flies at an astonishing array of species – red snappers, blue-spangled emperors, milkfish, yellow-tailed rock cod, triggerfish that scooted down boltholes when hooked. Although we came upon a few bonefish on Ray Charles flat (the joke being that even the singer would have been able to see fish there, so bright was the sand), I was never in the right part of the lagoon for any numbers – one definition of 'habitat' is a place where a particular species was last week – and some of the flats were as bare as Old Mother Hubbard's cupboard. Other groups fared better, and I regret my frustration became evident: 'Fishing is cruel,' shrugged Olivier. One morning, the sunshine coming down like poured syrup, six of us on foot were wading the centre of the atoll when, with almost no notice, heavy clouds shuddered up from the horizon, a sharp wind began to hustle the rising tide so that waves sloshed over our backs, and the recently serene surface of the lagoon became a riot of wrecked light. There was no dry land for miles, and suddenly there was the dark applause of a torrential cold rain, causing near darkness at noon. The rescue Zodiac could not at first find us even with its spotlights, by which time the tide was reaching my armpits. On Cosmoledo we never again felt like mere tourists.

High-end tourism was precisely what we experienced a few years later when Helen and I joined a family that had taken over the entire Seychellois resort of Alphonse Island for a week of sport and sunshine that began with a private charter and at times resembled a scene from some J. G. Ballard novel, with a far-flung group of sybarites beneath some palm trees ignoring the entropy of this world. The main fishing here is on nearby St François atoll, 10,000 acres of flats nicknamed 'God's Aquarium' for the clarity of water and spectrum of species – over sixty have been caught here on the fly, and you could rack up a dozen in a day if you wished, some of which even the guides had difficulty identifying. As well as several small Indo-Pacific permit, I tried for vermillion grouper, pompano, emperor sweetlips and even threw at a harlequin shuffle of parrotfish; I tried repeatedly to pick a fight with a whopping great milk (*Chanos chanos*), which refused to be provoked even by my drifted plankton pattern; on the coral 'finger flats' we gingerly stalked three different types of trigger, something of a cult quarry out there and infuriatingly inclined to nibble at your crab's artificial legs, then contemptuously whirr off elsewhere. Gorgeous bluefin trevally streaked in for speedily retrieved flies and there were also a few giants, though nothing like the population on Cosmo. In fact, only two-thirds of a geet was caught by the entire party all week.

While sharing a forty-bone morning with my skiff partner Robbie – the north-west trade had blown away two days of sultry weather and a brisk tide was flushing cooler water over the sand – I saw an indistinct grey shape pushing a rapid vee towards the drop-off and with the help of a following breeze I made a long cast that for once delivered my Mantis Shrimp right into what I considered his bone zone. I was swiftly taken, but immediately recognised the inimitable power-surge of a trevally (a small giant) which zipped off into the deep water – there was an unseemly convulsion, Rudi the guide shouted

'Shark!' and I reeled in just the top half of my geet, its eye still twitching (the section we had left would have weighed about 10 pounds, and we counted it as caught but discarded). That afternoon it was our turn aboard the billfishing boat, and I have already described the fiasco when my fly-line jammed on a bolting sailfish, thus depriving me of a unique Bills, Bones and Geet accolade. 'The dumplings in your dreams', runs a doleful Yiddish proverb, 'are dreams, not dumplings.'

Despite these many other temptations, it was the bone-fish that continued to detain me – after all, these are the creatures whose flickering images seem to obsess me when I am elsewhere in the world, strap-hanging on the Tube, or biting a ballpoint at my desk. When conditions are perfect, Alphonse bones are described as flowing onto the flats like a river, though I suppose the collective noun ought strictly to be an ossuary. Helen, who, settled with her book beneath a parasol, normally ignores the busy drama of the flats and my intermittent Captain Haddock soundtrack, decided to try her hand again at bonefishing here (though I had only just forgiven her for the Deep Water Cay debacle twenty-six years previously), and leisurefully wading Mullet Bay she accounted for seven nice boneys plus a brassy trevally. 'You should let your fly sink, you know,' she informed me, 'because they feed on the bottom.' 'I sure wish I had realised that all those years ago, sweetlips.' Yousef the guide even took us to a sandy promontory where tame bonefish jostled up beside the skiff when he cut the motor, and you could hand-feed them the remains of your picnic – nowhere else in the world have I had my dream fish nibble and suck my fingers like a puppy.

At Single Palm on our final morning, by ten o'clock it was incineratingly hot, but the visibility was exquisitely good and there were bonefish advancing like ethereal elvish armies glimmering in the morning sunlight; for two hours they processed in such impressive numbers that some even swam between my legs, and I only gave up casting my little Pillow Talk when I was dizzy from heat exhaustion. Lunch was taken at anchor on a coral bar where giraffe-marked rays were slithering through the streaming shallows, and in the distance the greenish sea melded to the blues of the horizon. That night, as we changed for the last supper, a spectacular electrical storm blew up – lightning threaded the heavens, and there came the banshee cries of hundreds of frigate birds expelled from their roost above our villa, screeching away into the forked and riven sky.

15

OF EATING AND DRINKING

'This dish of meat is too good for any but Anglers or
very honest men.'

Izaak Walton, *The Compleat Angler* (1676)

Over the years it has become a bit awkward, but, with a few
exceptions, I really don't like eating fish. My aversion stems
from childhood, and the weekly ordeal of a dish known as
Smelly William (some sort of black-skinned fillet poached in
oily milk), and at prep school I dreaded Fridays with its pervasive piscine miasma that leaked through the kitchen hatch.
Since I have difficulty tucking into anything with fins, I kill
relatively few fish – and have ignored the nursery maxim that
their flesh is good for the brain. I am fascinated, however, by
the way the fish has become virtually a symbolic foodstuff,
like honey, salt or corn.

In the ancient world fish were widely venerated, and exceptional specimens were sought-after as a badge of social status. The cult of fish epicurism or 'opsophagy' was a prominent aspect of the decadence and exhibitionism displayed by the governing elite of the Roman Republic, reaching its height around the time of Cicero (who became a consul in 63 BC). This striving for prestige by acquiring the novel or esoteric became as much a fetish as the later, seventeenth-century craze for Dutch bulbs (tulipomania) or our modern vogue for fancy koi carp. Choice fish were presented as gifts: especially prized were the *rhombus* (turbot), *murena* (marine eel) and *mullus* (the red mullet, but presumably not for thrusting up any adulterers' fundaments). Coastal villas boasted ornamental ponds (*piscinae*) with intricate systems of cisterns and aqueducts, sweet and salt, where particular prodigies could be admired. In his rambling 1854 study *Prose Halieutics*, the Revd David Badham records how Hortensius the orator, having never shed a tear at the death of seven wives, wept uncontrollably when his pet lamprey expired (there was a lively trade in lampreys and eels, considered – with their phallic associations – to be aphrodisiacs).

These trophies also supplied banquet dishes, and certain species would be brought alive to table in glass containers, so their colours could be admired as they changed in death. Recipes became extravagant: the epicure Vitellius devised one known as the Shield of Minerva, which combined peacock brains, flamingo tongues and the livers of many wrasse (*scarus*). Plutocratic gourmets greedily sought out

316

various refined fishy-essence sauces, such as *garum* and *liquamen*, equivalents of our Asian nam pla, and ancestors to Worcestershire sauce. A study of human coprolites suggests that despite the introduction of sanitation across the Roman world, the popularity of *garum* may have been responsible for the spread of parasitic fish-worm infestations, since latrine water was regularly used to fertilise crops. This pescatarian craze seems to have declined by the end of the first century AD, though its modern counterpart persists in ichthyocentric cultures like Japan: in 2019, a single bluefin tuna was sold in the Tokyo market for £2.5 million.

Right up until the discovery of refrigeration – probably introduced in 1786 by Alexander Dalrymple, who had observed Chinese junks packed with snow – fresh fish in Britain remained largely the privilege of the ruling classes, though account books show considerable quantities of smoked and dried seafood being consumed away from coastal areas. Pickling, salting and preserving with powdered charcoal allowed fish flesh to be transported inland, and hard, wind-dried 'stockfish' provisioned generations of sailors, pilgrims and other travellers. In Europe, where many people live a long way from the sea, the practice of storing fish alive was more common: carp were kept in damp moss hanging from cellar slings, for instance, or transported in panniers, their mouths plugged with bread soaked in alcohol. For centuries, fish markets were notorious for sharp practices (faking freshness, or deliberately passing off species) and for their colourful vocabulary. Billingsgate developed its own whole language of invective.

One reason such fishy supplies were so necessary is that the calendar of the mediaeval church forbade the eating of meat for almost a third of the entire year – and not just during Lent, though this was a convenient time to fast, since stores were anyway low after the winter. Unlike the stricter Jewish dietary proscriptions (fish had to have scales and fins), the Catholic rules were flexible enough to include occasional beavers, seals and sea-birds. Fish were nonetheless a valuable commodity, and it is no coincidence that many cathedrals were built on rivers where runs of migratory fish could be harvested (Ely was said to have been constructed on the revenue from eel sales). Weirs and wheels and various 'fixed engines' abounded – a dam fitted with a net was a 'kiddel', giving us the phrase 'kettle of fish' – and at one time salmon were numerous enough that Henry III had his pet polar bear let out from the Tower of London on a chain to catch its own Thames supper. The canard persists about apprentices – from Axminster to Stirling – protesting about being fed salmon too often, though this is likely to have concerned cheapskate employers supplying out-of-season flesh, such as kelts. By Tudor times, however, the rivers of England had become so plundered and obstructed that this abundance was drastically diminishing, which made the discovery of the New World – that second Eden, with its teeming shoals of sturgeon, shad, smelt and salmon – all the more fortuitous, and supposedly another sign of God's benevolence.

Evidence of pisciculture figures rather rarely in the Domesday Book, but by the thirteenth century the breeding of freshwater fish had become a serious study in monastic circles, and in some Continental houses there was even a specific post of *Fischmeister* to supervise this crucial part of their economy. Carp were popular for raising in stew ponds, but gradually, as fish epicurism among the nobility again became a mark of status, other species became desirable. Pike, for instance, could fetch ten times the price of turbot – they are notoriously tricky to cultivate, and hard to catch on the angle because it requires a wire line (in the Middle Ages this was costly and had to be commissioned from an armourer). Fishing – unlike hunting – was never prohibited by canon law, so it was for centuries associated with the comfortable religious life. To serve up an esoxian dish, therefore, bespoke influence: in *The Canterbury Tales*, Chaucer's Franklin, who aspires to gen-trification, boasts 'many a breeme and many a luce [pike] in stewe', while in the poet's delightful ballad 'To Rosamounde', he describes himself as sloshing around in love like a 'pyk walwed in galauntyne' (a jellified sauce with breadcrumbs, typical of the piquant garnishes devised for fifteenth-century sophisticates).

The pike, a fish now seldom served in Britain, though still esteemed on the Continent, is one example of how certain fare came to be associated with almost totemic qualities because of rarity value: they tended to be the prerogative of the elite, and thus were presumed intrinsically superior. A fine anecdote from the antiquarian John Aubrey's inimitable *Brief Lives* concerns the ecclesiastic George Abbot (1562–1633), whose mother, 'a poor cloth-worker's wife in Guilford', while pregnant dreamed that if she ate a jack pike her child would achieve greatness. The very next morning, while fetching water, she caught one in her pail (so likely!) and duly served up this prize at the infant's christening. The biographer summarises the rest of Abbot's life thus: 'he was bred up a scholar in the town, and by degrees came to be Arch Bishop of Canterbury' – a masterpiece of compression and causality, omitting such details as the subject shooting a gamekeeper with his crossbow and, on another occasion, being a notorious martinet, sending to prison 140 Oxford undergraduates for not removing their hats in his presence. And all thanks to predestination, plus the power of pike flesh.

Not every mediaeval dietician was convinced that fish was wholly beneficial, however, some of them warning against its richness unbalancing the bodily humours and provoking phlegm or the flux. Excess was unlikely to afflict the working man, but every schoolboy 'knows' that lampreys did for Henry I (it may not have been a simple surfeit: there are two threads in their backs which render certain lampreys toxic). It was popularly believed, too, that eating salmon could cause leprosy, since dried fish were standard travelling rations and so many returning Crusaders had contracted the disease – though in fact it seems to have reached Britain in the fifth century through infected red squirrels.

There is often an intricate link between food and nationalism, and traditional dishes feature strongly in our sense of identity and belonging. This can be especially true for émigré or expatriate communities, and naturally furnishes potent links with childhood, as well as reflecting all types of fad and taboo.

Although recently supplanted by chicken tikka masala, fish and chips was for a century or so regarded as the national dish of England. It is in fact a serendipitous immigrant marriage between pieces of battered seafood from Jewish cuisine and pauper's street food from Victorian times, traditionally wrapped in yesterday's newsprint spawned by us humble, jobbing hacks and scribes. I can just about handle this combination (particularly if there's a chance of goujons of sole on offer), but many of the delicacies from my adopted homeland of Scotland continue to appal me, including cullen skink (which looks about as appetising as a Shanghai spittoon), and the feather-boned, two-faced kipper. One hellacious Hebridean recipe for 'sour skate' involves soaking the fish for a day in horse dung before boiling ('a verra deleecious dish', a crofter assured the Victorian author Cornwall Simeon), while from Shetland comes a type of cod's head stuffed with oatmeal and haddock liver, nicely named Krappit Head.

Scandinavian fare appears to me nightmarish, with its myriad rollmops and dried cod. In Sweden there is a version of fermented herring called *Surströmming*, and it's said when the barrels are first opened its exhalation is so pungent that birds flying overhead drop from the sky. Similarly, the vicar of St Ives in 1870 claimed the smell of pilchards was 'so terrific as to stop the church clock'. Icelanders have been fighting cod wars since the fourteenth century: fish and geothermal energy are their twin staples. When it was still a Catholic island, its priests were granted a papal dispensation to consecrate dried fish as communion wafers in times of cereal shortages. Icelanders still delight in *Hardfiskur* – savoury fish flakes that W. H. Auden reckoned tasted like toenails, though one would prefer not to learn how he knew. He also described *hákarl* as tasting like boot polish. This delicacy looks innocuous enough – a pale cube offered on a cocktail stick – but it is in fact the putrefied flesh of a Greenland shark (a deepwater species that can live for three hundred years), and the decomposition process involves months of burial in the sand with periodic saturations of urine. To me, it tastes like a core sample from Satan's long-drop (which, incidentally, I am certain I discovered on the Tibetan side of Everest's first base camp). Like many people, I have an irrational suspicion of all shark meat, as the creature itself is uncanny; the flesh of some contains the nerve agent trimethylamine, which can actually be narcotic and induce a coma.

A gourmet tourist on a flying visit to Boston, Massachusetts, once wanted to sample its famous fresh fish speciality, scrod. He hailed a cab. 'Say, do you know some good place nearby where I can get scrod?' 'Sure, bud,' replied the driver. 'But that's the first time in twenty years I heard anyone use the passive voice of the pluperfect subjunctive.'

On my travels with rod and line, I have encountered my fair share of outré 'specials': in search of carp up a Thai *khlong* we stopped at a floating eatery and were served bat's blood soup followed by a casseroled pig's pizzle, but I have managed to avoid Rocky Mountain oysters, live baby dormice dipped in honey and the durian – that fruit so repulsive to smell that it has to be cut up outdoors, and is reputed to be redolent of spring onions over which someone gorging on chocolate has recently regurgitated. Entomophagy is nowadays being promoted as a source of sustainable protein, and I can recommend Calvin W. Schwabe's enterprising volume *Unmentionable Cuisine* (1979), which describes white ant pie from Zanzibar and a succinct recipe from Laos for 'Mang Par': 'Boil dragonfly nymphs. Eat them' – which might just do for a luncheon beside some Hampshire trout lake, accompanied by a bottle of pink Sancerre.

Human culture is stuffed full of bizarre comestibles. Even by Victorian standards, Frank Buckland, the roly-poly bearded surgeon and naturalist, was an eccentric showman. His father William was a geologist and canon of Christ Church, a zoophage whose ambition was to eat his way through the animal kingdom. Once, at an Oxfordshire dinner party, he is alleged to have devoured the desiccated heart of the Sun King, preserved as a relic. He introduced Frank to experimental feasts during which, in the interests of expanding the diet of the labouring classes, they sampled everything from horse jelly to boa constrictor. Young Buckland eventually tried roasted panther, sea slug soup and rhinoceros pie, but decided the bluebottle and mole were least palatable. A keen salmon angler, he developed a lifetime obsession with fish, and in Bond Street once tried to strap a 200-pound sturgeon to the roof of a hansom cab. In his house by London's Regent Park (I used to pass the address on my way to school), he installed a fish hatchery, and was instrumental in the translocation of salmonid ova to the Antipodes. Despite his foibles and occasionally slapdash science, Frank was a serious investigator of methods of farming fish and other marine creatures, and an early whistle-blower about the decline of marine-fish stocks. In 1867 he was appointed Her Majesty's Inspector of Fisheries. An entrenched opponent of Darwinism, Buckland believed the common dogfish was one of the more perfect designs in Creation.

But his outlandish dishes are as nothing set alongside *The Futurist Cookbook* published in 1932 by the Italian proto-Fascist poet Filippo Tommaso Marinetti. At some of his 'Extremist Banquets', no food at all was served, but aromas were broadcast over guests instead; the author had developed a madcap concept of *Conprofumo* (interesting to me, as my surname means 'perfume' in Italian) whereby certain foods enjoyed an olfactory affinity, such as roses and cooked potatoes. His fish recipes for actual dishes include Alaska salmon in the rays of the sun with Mars sauce; Immortal Trout (fish wrapped in thin slices of liver, paired with a devil's nightcap of a cocktail comprising grappa, gin and anise liqueur and a square of anchovy paste wrapped in a wafer); the ominous Black Shirt snack (a cod cutlet flambéed in rum); marinated eel stuffed with frozen minestrone; a Holy Supper bowl of fish fillets buried in whipped cream and zabaglione; and Drum Roll of Colonial Fish (a poached mullet stuffed with date jam and eaten to the sound of military percussion). Escoffier would have been turning in his gravy.

*

The Balmoralisation of the Highlands resulted from Sir Robert Gordon choking on a breakfast fishbone in 1847, making way for Prince Albert to acquire the house. Breakfast is definitely the time of day when I most abominate fish-eating, especially kedgeree steaming on the hotplate, or my wife's unfortunate penchant for haddock, and kippers. I have invented a maxim that serving fish before a day's fishing brings bad luck, and the nearest I ever come is an infinitesimal layer of Gentleman's Relish on the toast before golden scrambled eggs are scooped upon it, offset by a great dish of malty Irish blend tea.

Picnics are like parachutes: to avoid surprises, you should always pack your own. Bapped offcuts and a Thermos of some hotel's dishwater soup are most unwelcome, and unless we are going to enjoy an afternoon's farctate snooze I prefer to travel light with a simple 'piece' in my pocket (buttered oatcake, cheddar, *saucisson sec*, Kendal Mint Cake): besides, normal lunchtimes often coincide with fish-taking times, so it can prove counterproductive to stop. However, even I would concede that the best context within which to eat a fish is at the water's side itself – cleaned in the lake, baked in a brushwood fire with a 'hazelnut' of butter and eaten with picky fingers, a finnock or brownie can be, in the words of Father Walton, 'excellent good'. More than once, on a Cuban mothership, I have enjoyed tuna sashimi from yellowfins we caught less than an hour previously: that sort of association closes the whole circuit of angling in a quintessential, atavistic way.

A sporting day's provisioning can be as necessary a part

of the ritual as dressing for the kill. Walton's Piscator on the Lea enjoys 'a brave breakfast with a piece of powdered beef and a radish or two', while Zoffany depicted the Garrick family at an al fresco tea during which a chap in a tricorn busily wields his rod. Edwardian Patrick Chalmers, ever unhurried around his beloved Thames, contemplates 'A cigar among the roses and possibly a tankard of cool, brown ale about eleven o'clock'. In Chile in 1937, Negley Farson records a memorable interlude with cold, boiled rainbow trout and a bottle of hock. Hemingway (who meticulously chronicles what his characters eat) can himself be found on the Rhine in 1922 happily finishing a paper bag full of cherries. Aboard the *Pilar* off Cuba, he would make lunch from alligator pears and vinaigrette, washed down with those infamous cocktails in a metal bucket. Which brings us to the historic affinity between fishing and drinking.

It is said that the benefits of secondary fermentation were discovered inadvertently by an angler. In the days of Bloody Mary – the temperamental Catholic monarch, not the cocktail Papa claimed to have invented – a worthy Protestant divine named Alexander Nowell was fishing a favoured swim at Battersea when he was warned that royal agents were coming for him, so he hastily buried his lunch bag in a bankside hole and slipped away. In the event, he was constrained to seek safety in Strasbourg, and it was not until some years later, under Good Queen Bess, that he returned to the spot, dug up his stoppered bottle of ale and found it had improved immeasurably. He never made a groat from this boon to civilisation, but later became Dean of St Paul's and, from his portrait now hanging in Brasenose College, Oxford, he looks contented enough, with his open packet of hooks. He lived to the then almost incredible age of ninety-five, and Walton (who himself almost made it to ninety) remarks, 'It is said that Angling and temperance were great causes of these blessings.' That sounds like Anglican propaganda to me.

Another of mankind's more ancient, widespread and earnestly pursued activities, drinking enjoys a historic symbiosis with all forms of hunting, and *Homo horizontalis* surely evolved even before the dawn of civilisation, gruntling with delight at the chance transformation of stored fruit juice, liquidly celebrating good fortune with spear and club, eventually bestowing upon alcohol (and other intoxicants) the mysteries of religious experience and other sacramental qualities that survive in certain rituals even today. Our own piscatorial culture remains nicely irrigated with various libations to appease the river gods, celebrations to wet the head of a triumph (as pleasurable as any post-coital cigarette) and the inevitable hat-stretcher of a tot to accompany the tall tale. 'It's a puir story that disna' end wi' a dram,' as they say on Tweedside. We may resort to a jolt of Auld Shibboleth as a febrifuge to chase chills after a ducking, or the consolation of John Barleycorn following some fiasco, and there are numerous other excuses for toasts, sundowners, nightcaps, chasers, shots, 'wee drappies' and cordials various.

In assorted lush dives and low joints on my travels with

rod and line I have tried many variants of neck-oil, bust-head, squirrel dew, choke dog and mountain brandy, from Swahilian tembo to a Connemara *poitín* that smoked slightly as you uncorked it. Maybe the worst was a bottle of ersatz Kashmiri 'Scotch' that made your uvula shrivel and brought a rictus of relief when your mouthful was safely down the red lane (Faulkner's term for hooch like that was 'Jefferson hair tonic'). With age, I have come to be a little less ombibulous – beer seems excessively diuretic, brandy makes me walk like Andy Pandy, and I have had to foreswear previous cocktail delights such as the Depth Bomb (a shot glass of bourbon lowered into your ale) or the Percy Special (Scotch and cherry brandy) as they give me a mighty headache as if Pallas Athena was about to burst, fully armiferous, from my frontal lobes. For some it is the subway to oblivion, but the general idea of inebriation is to render pleasurable the moment, and our language resonates with several thousand terms for its spectrum of effects – gold-headed, sherbetty, grape-shot, elephant's trunk, whipcat, poddy, shickered and cherubimical (again, I am listing slightly). The king is his cousin, he has lost his shoe, there is a guest in the attic – he is down with the fish . . .

Come with me now – it's just after one, on a dreich March day outside the Grog Palace, a fishing hut on Royal Deeside where a party of rods is on their annual safari after a rare spring salmon. Peer through the bleary pane, where the chaps have congregated for a sharpener (their ladies have gone 'tweed-ing' in Ballater). Billy the gillie in his herringbone breeks is prodding the stove back into life, and Al Cockermouth – wine merchant, team leader and *arbiter bibendi* – is constructing his trademark cocktail, the Red Snapper (rhubarb gin and house Prosecco), the luncheon brace of Gigondas magnums already breathing on the mantelpiece. There's Tom Hernshaw, Norfolk JP ('ever hear a jay pee? they're damn quiet about it'), who is peeling off his neoprenes, turtle-bellied in his Orvis tartan plaid shirt and peacock pattern braces. Guy (Viscount Sparling) has been squiring Tom's daughter Jo-Jo for a year now, and everyone is hoping for some kind of announcement soon; he is feeding broken Bath Oliver biscuits to Pucci, his mutinous springer. The fourth rod is media lawyer Nigel Paternoster, a golfing chum from Kent ('not much of a fisherman, I'm afraid'), who is disconsolately searching for a phone signal. Young Sparling has pulled out a kelt, the only fish of the morning, so there's nothing particular to celebrate, except for bonhomie, good fellowship, time away from the office, a sense of reunion. '*Skål!*' says Tom. 'Sluice your dominoes! *Sláinte*, Billy! After all, there's more to this business of fishing than just catching fish.'

All harmless enough fun, but strong waters burn deep and don't always mix with fishing. Accidents can befall the befuddled, over-refreshed with a snootful of the Red Infuriator, and you don't want to tumble base over apex into the bouldery, freezing stream because you are wading like a Fimble Fowl (Edward Lear's bird with a corkscrew leg). Sorrows are all that need drowning. You will also do well to watch your step next morning, should you still be testing positive for Johnnie Walker and have a 'tartan heid' full of hot clinker, plus a conviction that during the night the entire Algerian army has marched barefoot over your tongue. In a chapter of recipes such as this there should be a few formulae to assuage hangovers: Pliny the Unreliable swore by owls' eggs, Coleridge relied on laudanum with his fry-up, while the chemist Robert Boyle advocated putting hemlock in your socks. I recommend a West Country remedy, if you can keep it down: Thunder and Lightning requires two spoons, one of clotted cream and the other of honey in the comb – you take alternate nibbles from both. On second thoughts, Father Walton's combination of angling and temperance might truly be the better bet. That stylish and sensible storyteller V. S. Pritchett advised the key to longevity was to drink like a chimney, and smoke like a fish.

The cantankerous Kingsley Amis was an expert on booze, and wrote an insightful monograph (*On Drink*, 1972) which distinguished between the 'physical' hangover and its 'metaphysical' aspects; he once asked a fellow novelist, who had been complaining about feeling down in the dumps, 'My dear chap – are you sure you're *drinking* enough?' The intersection between writing and drinking makes for a long, painful saga in itself, as there have been potomaniacs aplenty at the typewriter, and of course the notion that alcohol promotes inspiration is perilously undermined by the fact that it is a depressive, and unlikely to unfetter the bardic tongue or act as the fuel for fiction in any kind of long-term capacity (it is seldom really a fish ladder to help you up those daunting linguistic Falls). Nonetheless, some writers resort to all manner of stimulants to sustain their momentum and fend off the hobgoblins of lassitude, uncertainty and despair – Disraeli composed his novels in evening dress, while spooning up champagne jelly; Schiller liked to invigorate his travails by sniffing a drawer full of rotten apples; and Balzac binged on up to fifty cups of coffee a day. On Mount Parnassus there may have been two peaks, one to Apollo the god of poetry, the other to Dionysus, but the latter is the one casting the longer shadow. Is there, though, a fundamental similarity between the experience of drinking and fishing?

It is tempting to describe them both as potentially addictive – the click of that alimentary orgasm corresponding to the ecstasy of a thrilling hookset. Both encourage you to look beneath the surface of things; and their pleasurable effects – subtle gradations, which are subjective and vary between individuals – can alter your perception, and give the impression (however momentary or illusory) of unlocking the little doors of the universe. Some people call this escapism, but I've never quite understood what's wrong with that, anyway. It's not just that a beaker of moose juice seems to realign the cosmos, or that a fishological moment of sublime satisfaction on the water acts as a corrective lens; I feel there may be something obliquely mystical at work here, that sensation that unspecified elements of your world are slipping into place. The Harvard philosopher William James, brother of novelist Henry, whose work influenced the founders of Alcoholics Anonymous, wrote about the seductive effects of inebriation:

'It brings its votary from the chill periphery of things to the radiant core. It makes him for the moment one with truth' (*The Varieties of Religious Experience*, 1902). I'd say that describes an identical experience at the heart of angling.

Knocking it back never appealed to me as a teenager until, on my gap-year travels through India, I had my first proper night on the Scotch in a maharajah's garden spread with rugs and lit by flambeaux. Asked what I would like to drink from the ornamental trolley being trundled around by a sleek attendant, I panicked and chose the first bottle I could readily identify, which was Ballantine's whisky. There followed a series of what the topers of the Raj era would have called 'chota', or, rather, 'burra pegs' (I've always admired that character in *Scoop* who wanted a 'chota mallet'), and for some years I stuck to such an evening diet – indeed, I virtually dipped my quill in it, and with a young journalist's kamikaze appetite I also smoked two packets of cigarettes a day, a habit that would now set me back some ten grand a year (I admit they were only Silk Cut: I had long since given up trying to impress girls with my untipped Gauloises). These days I have moderated things marginally, but am still fond of a single pot-still Irish whiskey named Writers Tears (no apostrophe). Intrigued as I am by the relationship between creative people and drink, nobody much wants to end up like Dylan Thomas: his final words, in Room 205 of New York's celebrated Hotel Chelsea, were said to have been, 'I've had eighteen straight whiskies ... I think that's the record.' In fact, more poignantly, he muttered to his female companion, 'I love you, but I'm alone.'

An unsurprisingly large number of *angling* writers have been no stranger to the toddy tumbler. One imagines that Walton himself was something of a sobersides, but there's little doubting the tastes of his co-author Charles Cotton, a carouser who began one ode promisingly enough: 'Come, let us drink away the time,/ A pox upon this pelting rhyme.' There was a healthy fashion for roistering piscatorial verse and Anacreontic drinking songs in the Restoration period, though angling as a pastime with Arcadian overtones and an image of seclusion has never quite had its equivalents of the rowdier equestrian pursuits, with those Regency bucks and delinquent hunting fanatics such as the Shropshire squire 'Mad Jack' Mytton, spendthrift Hussar and MP for Shrewsbury, who drank seven bottles of port a day, rode a pet bear, threw his fiancée's lapdog on the fire and, pursued to France by debt collectors, died of brandy in 1834.

There has been a line-up of quietly heroic halieutical boozers, however. Van Campen Heilner once missed an entire Cuban tarpon season in the 1930s because he was so taken with the delights of Havana's bars. Though impressively rich, to appease his mother he briefly agreed to work as a panellist on a weekly radio programme, *The Rod and Gun Club of the Air*. He arrived so crapulent one morning that when he attempted to answer host Dave Newell's question, 'Van, don't you think the bonefish is just about the fastest fish in the sea?', he was sick all over the microphone – whereupon his colleague improvised masterfully: 'That's right, Van. The barracuda may be faster in the dash, but he can't keep up with the bonefish over a quarter mile course.' Papa, too, was legendary for (as he phrased it) 'bearing down heavy on the sauce', and became pugnacious when shellacked, sometimes even using as punchbags the corpses of fish hanging on the dock. Negley Farson, another hard-living American genius, was frequently as drunk as a Gosport fiddler, and had to be admitted to an asylum in Berne. Pulitzer Prize-winner Robert Lowell was a serious alcoholic, and often fell in while angling; his poem 'The Drunken Fisherman' (published in 1946) is arguably the scariest joint evocation of these twin subjects.

In *A Shropshire Lad* (1896), A. E. Housman famously claimed, of alcohol, 'And malt does more than Milton can/ To justify God's ways to man.' In 1922, the Cornish novelist and critic Arthur Quiller-Couch asserted, about teetotalism and writing: 'A total abstainer is in the nature of things imperfectly equipped for high literature,' and this was glossed by a London journalist as, 'A good stiff scotch splash, and you cannot think what a merry and bright little trifle *Paradise Lost* seems.' Is drink, in fact, an integral part of our pursuit of Paradise?

W. C. Fields was in no two minds about where happiness was to be found: asked why he never drank water, he replied, 'Fish fuck in it.'

16

THE MOJITO COAST

'Yo soy un hombre sincero,/ de donde creca la palma.'
'I am an honest man,/ from where the palm
tree grows.'

José Martí, from *Versos sencillos* (Simple verses) (1891)

When the inflight movie is in black and white and the drinks
trolley runs dry before it reaches you, it's a fair bet the state
airline is heading for a time-warp destination afflicted with
chronic shortages. As we bank over the capital most of the
buildings have been darkened by power cuts – this is April
1997, and Cuba is struggling in its Special Period following the
disintegration of the Soviet Union, which had been subsidis-
ing its economy with an annual six billion dollars. Dr Castro
is urging his people to fashion shoes from banana leaves, even

a surgeon earns only twenty dollars a month (with a bar of soap thrown in), and others are reduced to washing themselves with Chinese toothpaste. In the absence of any cheese the pizza sellers are topping off their handiwork with melted condoms (I wouldn't really recommend the pepperoni, either).

A place of rust and pollution, falling masonry and dripping standpipes, La Habana nevertheless wears a ramshackle beauty – glamorous and pockmarked as a faded courtesan. In its dissolute heyday under Batista (when the lascivious Graham Greene enjoyed the live animal acts at the Teatro Shanghai) it could boast 270 brothels and numerous opium dens, and even now it is pre-eminently a city of the night. Cruising the Malecon's seafront boulevard in a 1950s Pontiac on our first ever evening, we pass men listlessly fishing its oily waters (condoms here inflated as floats) and head downtown to the warren of cobbled side streets where tobacco smoke and bebop music leak from every bar, the limestone facades are salt-bitten, and the air carries the thin reek of unsuccessful plumbing. At trestle tables, grouper-jawed European males with an appetite for 'city picnics' eyeball the Spandexed girls, zebra-striped as reef fish shimmying through the submarine gloom. Dancers sway in shoals. With a bonefish pucker I slurp on my cocktail straw, eyes as glassy and startled as my favourite quarry. Already I am adrift on a Cuban tide, and far from home.

Despite oodles of forbidden fruit, the capital itself may be more Babylon than Paradise – but there is a necklace of some six hundred islets off the south coast which is a kind of accidental Eden. Driving through the fertile central provinces of Cuba you see walnut groves, cattle ranches, russet plough, frangipani and cane fields. With no more spares, Russian machinery languishes on every state farm. In places it looks like Devon, but with turkey buzzards and numerous signs proclaiming *'Socialismo o Muerte'*, which I suppose is more Islingtonian. From the shoddy port of Jucaro, which feels like the coccyx of Cuba, you embark on a four-hour judder aboard the antediluvian cruiser *Sol*, and there, lodged in a mangrove channel, is *La Tortuga*, a weather-beaten double-decker steel 'boatel' which will be your rickety haven for a week of mixed fishing in hundreds of square miles of nature reserve. This might not quite be *terra incognita*, but back then the Jardines de la Reina had not seen tourists in a long while. Columbus reached Cuba via these uninhabited cays in 1492, and later reported, 'This country is the most beautiful that human eyes have ever seen.'

It was only the second year of exploratory fly-fishing here, and from the start there was that rare sense of discovery. Guides merely had rowing boats with outboards (the cowlings of which were as dented as gladiators' helmets) and whittled branches for push-poles. As we chugged across the deeps – azure scumbled with cobalt – towards distant flats, needlefish spritzed away from our wake, minuscule jellyfish dimpled the surface like reverse rainfall, there was the seasonal sight of enthusiastically copulating sea turtles from which the 'floatel' gets her name. At the tiller was Pedro Maria Peleaz, the guide whom you wouldn't book as entertainer for your kiddies' next fiesta, but a man whose eyesight and instincts have done me proud over twenty years. Before conservation measures prohibited it, this former lobster trapper supplemented his diet with a slingshot that brought down everything from pelican to roseate spoonbill – any target with a wingbeat, though even he drew the line at the foul covens of cormorants that descend from vast mangrove roosts and splatter their feculent way across the shallows, scattering every fish into the next parish. He was also fond of *hutia*, an arboreal rodent and Cuba's only surviving native mammal, said to be scrumptious when potroasted with honey, and a popular snack for saltwater crocs. Occasionally Pedrito would clamber through the *mangli* and retrieve one to exhibit, liberating it with evident reluctance. Catch and release seemed anathema to any citizen surviving on the ration system of the *jefe máximo*.

Though now they are rather considered a support act, it was the bonefish (*macabe*) that was then the main attraction, since they were legion and had been virtually undisturbed. Seldom have I come across examples that hit the fly with such conviction – at times they seemed as hungry as the citizens themselves. Favoured areas included Mariflores, where you would bounce chenille crab patterns off the mangrove stems at high tide, or Las Crucesitas where shimmering dorsals bear down on you in a flotilla of jostling sails, and Pedro urges, '*Mira*, Dayvee. Eleben o'clock, one carsse, quicky. *Mierda!*' as you fluff your presentation – bone fever is as infectious as yawning. In Cayo Lisa, the bonedogs seem somehow simply to vanish over the turtle grass as you close in for your shot – a phenomenon known as *macabe fantasma*. On the reef side at low tide, they may tail tantalisingly in the surfy slatch, as you wade along a coral substrate with the feel of shattered amphorae underfoot. Grallatorial herons mince along the shoreline and ignore you like wine waiters.

The Gardens of the Queen was one funky destination. Seeking the shade at lunchtime (*la hora del diablo*) you might repair to the Iguana Hotel, a ragged thatched pavilion where you can feed orange segments to the resident lizards as you cool your forehead with a welcome tin of Cristal beer. The variety of wildlife was exceptional and surprising. Pedro took us to a stingray nursery, where baby specimens were crawling like lice, and with typically deranged bravura he picked up several 2-footers by hand, even though the sting is said to be almost insupportably painful (the spear tip of Telegonus was armed with one, and thus he unwittingly slew his father Ulysses). The entire Jardines was like some piscatorial red-light area where, depending on your special inclinations, the possibilities seemed spoiling. If the flats fishing somehow palled, you might troll the nearby reef for grouper, wahoo or tunny. Our small party clocked up over twenty species that first week. There was aboard a venerable mainland bass guide named Feyo, who taught us to freeline sardine baits into ocean boltholes for snappers and jacks, expostulating whenever some tackle-tearer or other escaped, 'Aw, somnavabeetch!' He also supervised the nocturnal bait-fishing sessions for tarpon in a deep *boca*, which we attended with much onboard rum and cigar smoke, the eyes of each hooked fish an ember in the torchlight. My impression of this ostensibly benevolent grandfather changed slightly when I later learned he had assisted in the execution of hundreds of Batista supporters in a volunteer firing squad.

The first parties of fly-fishermen had been Italians, and their bonefish patterns were rudimentary. Over subsequent visits, we soon established success with Pop's Bitters, the McVay Gotcha, and especially Bob Veverka's leggy Mantis Shrimp, which the guides had never before seen and in seasons to come morphed into the ubiquitous Cuban Shrimp. But there were times when Mr and Mrs Macabe, out on their evening *paseo*, seemed unfussed by their choice of street-food titbits. Don Jon and I once came across a dirty blonde flat where the fish were thronging, and for several hours my ATH reel was repeatedly sizzling away. We decided to stop at forty-three because by then my forefinger was bleeding from the continual passage of gritty fly-line and some small Men in Grey Suits were starting to show up. Our flats clothes were coated in bonefish slime, which looks rather like jism. Sometimes only too much is enough (as John Updike observed about sex and money) and the same is perhaps true of certain moments in fishing. There was another bleeding finger that night, too, when my multi-talented friend Neil Patterson, lacking a plectrum for his borrowed guitar, played a marathon singalong until the strings were reddened, and to my own astonishment, despite numerous Cuba libres, I remembered in their entirety the lyrics to 'American Pie'.

*

We always take a little *après pêche* time in Havana, and over two decades the city has of course changed significantly: it ill behoves the dollar-rich tourist to sound nostalgic – a case of sugarcane romanticism – but it was more fun to visit before the Pepsification of the Pearl of the Antilles, though clearly for Cuban nationals, caught in a demoralised Can't Do culture, those were desperate times. In El Líder's dystopia there were three hundred prisons, an insidious network of informers, and a mail service that took a fortnight to deliver anything. He had all the island's Monopoly sets destroyed. There was so little lavatory paper that folks resorted to using the propaganda tabloid *Granma* instead – which says something about Marxist-Leninism, and conceivably the status of journalism in general. Private enterprise was prohibited but, despite the rigorous system of tourist apartheid and the regular thuggery of Castro's impeccably dressed 'love police', hustling and jockeying (*jineterismo*) was widespread, and at times the Havana street experience was like a sex and shopping novel, only without the shopping bit. (There was rumoured, however, to be a shoe store in Habana Vieja called something like Esmeralda – but don't go searching for it on my say-so – where, if the shopgirls couldn't find you any footwear to fit, they would take you upstairs to a kind of Cupid's gymnasium for a compensatory bout of what Oscar Wilde was pleased to call 'buccal onanism'; I never went in search of any souvenir soles, partly for fear that they might not deal in half-sizes.)

Havana remains the Cleopatra of capitals – seductive, not classically beautiful, and wistfully past her prime. I have been there now a dozen times, and one of the loveliest experiences I had was when I slipped into the back of a rehearsal room in the Gran Teatro (a typical mixture of the splendid and the beaten-up) and watched their National Ballet practising in an upstairs hall moted with lines of light. Though materially poor, Habañeros manage brio and pep despite their penury, largely through dance, gossip, satire and music. A favourite jazz box was the Cafe Gato Tuerto ('one-eyed cat'), an archetype of the dive Irving Berlin described, 'Where dark-eyed Stellas/ Light their fellers' panatellas' – and indeed it was there that I recalled the lurid nail-varnished whippings on my childhood fishing rod's dusky blank. Everywhere after dark there is much to catch the eye: one velvety night in Miramar we accessed by buzzer, speakeasy-style, a pop-up restaurant (*paladar*) where, instead of the usual menu of roadkill surrounded with black beans, we were served turtle steaks by a waitress in hot pants riding a skateboard. In the adjoining room, some *yanqui* spooks were discussing the anniversary of the Bay of Pigs. Alas, I had to send back the Chablis, which was fatally corked.

In May 2001, while on a nice journalistic jaunt, I joined a party of angling Brits, which included my brother Mark, a bearded judge (when Cubans learned he was a *juez* back home they tended to look uneasy), in the company of a delightful government guide – 'My name is My Lai,' she said, 'after the massacre.' I visited a rambling apartment just off O'Reilly, where I had covertly arranged an audience with a *babalayo*. A syncretic faith that fuses elements of Catholicism with

Yoruba animism, Santería (or rule of the saints) is an unofficial Cuban religion that has an estimated five million devotees, back then said to include the Castro brothers. Their highly respected priest figures – owing more than something to the role of African witch doctors – are called *babalayo*, and Arsenio Martini was considered exceptional. He does not normally see outsiders, but I had engineered an introduction.

I gifted him a bottle of *ron añejo*, crossed his palm with some greenbacks and watched him apparently fall into a trance. He scattered some dried divinatory knucklebones (*ékuele*) on his mat, lit a cheroot, opened a cupboard that contained a shrine festooned with chicken feathers and blood, then proceeded to tell my fortune. He said I suffered from high blood pressure (easy guess for a guy of my age), was seeking some good luck at sea (I had already explained in translation we were going to fish) and then announced I had recently lost a brother (this was true, and unnerving). For success in my watery endeavours, he advised, I must wash myself with fresh mangoes each morning – easier said than done in most Highland lodges.

Although there is an overarching deity, Santería is based on the worship of individual spirits (*orishas*) that act as your guardians, like Christian saints. Yemaya, the spirit of the ocean, is identified with the Blessed Virgin Mary, and her son Changó is the Zeus-like lightning god. According to Arsenio, he was my tutelary spirit, and I should wear a bead bracelet (conveniently available from his assistant) in the sacred colours of red and white. Changó is said to like bananas, and lurks in royal palms, and is responsible for their tops being destroyed by lightning. He is known as the most temperamental and fearsome of all the *orishas* and, as My Lai solemnly interpreted, during ritual dances he sends thunderbolts out of his scrotum. I duly acquired the bracelet, and wore it semi-religiously (maybe that was my mistake) until on Independence Day later that year, in the Kola Peninsula, I forgot to do so, and our helicopter took its dive.

Some words just sound euphonious, irrespective of their supposed meaning. James Joyce liked 'cuspidor', and Henry James reckoned 'summer afternoon' were the two most beautiful words in the English language (to some writers, 'The End' would also be contenders). My own favourites might include: glimmer, asphodel, melancholy, persimmon, mandolin, consolation and *palometa*. This last is the lovely, fluttery Spanish nickname for the permit, a fish to which 'elusive' is regularly attached like some Homeric epithet. For the swoffer, it is definitely one of the most desirable challenges with fins.

Trachinotus falcatus is the platinum-plated, slab-sided chief of the pompano clan. Whereas the boney is elegant, Mr P looks vaguely extra-terrestrial: he has a glaucous, domed head and a demeanour as impassive as Robocop's. His mouth, set in a permanent sneer, is lined with skin as leathery as a Prada clutch bag and fitted with molariform grinding plates he uses to terminate crustacea with extreme prejudice. The body wears a zoot-suit sheen, with a smudge of mustard around the anal fins, though in the water he glides like a bluish parabola marked out by slivers of black detail to the dorsal. The spectacle of one busily tailing (*coleando*) is a sickle in the salt. Robopermit can grow to an astounding 50 pounds, but every least one counts as a trophy.

Sculpted from moonlight, powerful, inscrutable and polyphobic, for wariness and stamina the *palometa* makes the *macabe* look like some clubbable asthmatic. The permit hunter must develop a different rhythm from the bonefisher, too: the mindset must be narrowly focused, you ignore other temptations, you pursue the lonelier course of a sharpshooter, acknowledging that you might be lucky to get a couple of shots a day. Time, tide, wind and light all need to conspire in your favour – ideally, perhaps, 2 to 5 feet of water on a spring tide after a moonless night, with the sea temperature above 73°F, a broken cloud ceiling, that slight breeze to mask your approach, and a happy hour in some place where the fish are enjoying a gourmet tour before returning to the deeps. The main trouble is that permit tend to spurn most artificials: they window-shop but seldom succumb to an impulse buy (Tom McGuane suggested that it was like attempting to bait a tiger with watermelons). Often they ignore you – but don't take it personally, they're just being permit. I have spooked them at a hundred yards merely by unhooking my fly from the keeper ring (*'espantos'* sighs the guide – 'spirits') but at other times, just as you are hunkered down in the skiff changing your lure, Mr P will suddenly materialise nearby through his permitty portal and loiter insolently like some small-time hood enjoying his smoke break at the intersection of Main Street and Eighth Avenue, only streaking off to safety once your tackle is ready and rigged. I believe they may even have evolved a sense of irony.

It is thought the first person to take a permit deliberately on an artificial fly was the luminary Joe Brooks at Content Keys, Florida, in 1951, using a bucktail streamer. Al McLane (who concluded this species had the disposition of a neurotic monk) managed just sixteen in four decades of trying. Lefty Kreh recorded hooking three out of the first three thousand he cast to. For pickiness and cussed unpredictability, Platinum Jack makes the Atlantic salmon look like low-hanging fruit. Your ersatz crustacean fly is often weighty, needing to descend through the water column like some critter in search of sanctuary, yet it must be delivered with delicacy right within the permit's orbit, often on a longish leader, thus risking a 'McCrab earring' for your lug if you mistime that cast. Those early patterns have been superseded by intricate shrimp or crab variants, and indeed a number of anglers have begun to specialise in bounty-hunting the permit to the exclusion of all other flats targets. Permit are so highly sensitive to smell that on occasions I have resorted to spraying the fly with an extract of crustacean DNA, but gave it up when one of the aerosols leaked with disastrous pungency into my hold baggage. At Cayo Largo, the *palometa* tend to browse in the wake of a sting-ray's magic carpet, and you throw directly at these satellites, using a fastish retrieve and one of Mauro Ginevri's ingenious, keel-beaded Avalon Special patterns. And sometimes There Is A God, even if his name is Changó.

So far it has been a slow day as Pedro pole-dances our way

down from Boca Grande. Don Jon and I have boated a brace of bones apiece, but now my attention is wandering; I'm scanning the sky for cloud sculptures and enjoying the badinage of my mucker when, 'Lookee, Dayvee,' squawks our guide, *'palometa! Grande.'* And there, shouldering its way up the sandflat, appears a single, formidable Robopermit on patrol. I am loaded for bone – 8-weight line, 12-pound tippet, a smallish Hula Shrimp on the business end – but there's no time to swap rods, and I make my presentation cast. For once, it falls soft and true, 3 feet ahead of the fish and, against all the rulebooks, he ambles over, tips up and snaffles it like a cranberry cocktail at some urologists' convention. Twice I strip-strike, giving him the Judas Kiss, he sprints off smoothly 100 yards downtide, pauses, takes out his bag of tricks, and heads rapidly out through the perilous stone gardens towards the distant Caymans.

I have hooked some permit before, but this is a big one. For thirty-five minutes we chase him in the skiff, motoring in spurts to keep a modicum of backing on the spool. My palate has turned chalky with nerves, but eventually he begins to flag, circles the boat belly up, and becomes ours. He cannot be revived, so we slay him for supper. On the spring balance he weighs 25 pounds. I gaze at him longingly, like some lovelorn department store commissionaire watching that girl from the cosmetics counter on her al fresco cigarette break. Although we are now on the cusp of a coveted Slam, Pedro with considerable chutzpah elects to stop for a wretched rice and beans luncheon.

Afterwards, he leads me inland, both of us wading nipple-deep through some glutinous gumbo of a backwater, at the end of which is a foetid, overhung creek with an open area equivalent to a medium-sized cat flap into which I am bidden to pitch a 2/0 tarpon streamer. Second try, a baby *sabalito* hits and, excruciatingly, I trout-strike. When I repeat this feat, I need no My Lai to interpret Pedrito's utterance. The third time, mercifully, a tarpon the size of a decent dace is hauled across the sludge and duly photographed. It is 2.27, and I have been Slammed. '*Coño*,' grins the guide with relief. It is his first Slam, too, and one of the finest fishological moments of my life. Truly, 'the afternoon knows what the morning never guessed'. I doubt there were two happier chaps on this sweet singing sphere.

To celebrate my serendipitous trifecta that night I delivered several self-adulatory speeches worthy of Mr Toad, and proceeded to get dog-sucking drunk. Back home, I commissioned a painted replica of my permit, carved from jelutong wood. I applied for the official IGFA Slam certificate, which now hangs framed in my tackle room like the accreditation of some questionable psychotherapist. There were ostentatious badges and chevrons to be sewn onto fishing shirts, guaranteed to irritate fellow sportsmen. As the hotelier Charles Ritz once remarked to Nick Lyons, 'This saltwater fly-fishing, it is for men with hard stomachs – like sex after lunch.' Now, that would be a certificate worth having.

Probably my favourite hotel in Havana is the spartan Ambos Mundos, where you can step straight off the street into the lobby bar and order your mid-morning mojito: the secret to this much travestied drink is excellent ice (by no means a given in Cuba) and to insist on the barman using *yerba buena* – tropical spearmint, its stems bruised. It was in Room 511 here, shutting himself in with a 12-pound ham and two stacks of blank paper, that Hemingway wrote, among other things, *For Whom the Bell Tolls*, a book that later inspired Castro and his guerrillas. One year, Esperanza (the golden-haired curator of his little museum) let me sit and type briefly at his old Royal machine. You can't go far in Cuba without coming across signs of Papa, who first visited in 1929 and soon became a familiar sight around town, in his blue gingham shirt and shorts. He remains part of the nation's folklore, even though, for copyright reasons, his works are not widely available there. In El Floridita, now a clip joint, though retaining a certain decadent air, he devised the Papa Doble (grapefruit juice, rum, emphatically no additional sugar) and one imagines it might have been entertaining to have been there on the evening he 'made a run of sixteen' such cocktails, though in truth Don Ernesto, despite back then being able to hold his booze impressively, was a bully and an egotist and a poor loser. He once said that he liked the island for 'both fishing and fucking' – perhaps not one of his more lapidary phrases, but it has the merit of honesty.

The pioneer of billfishing these Gulf Stream waters for sport, Papa bought a bespoke 38-foot motor cruiser (the *Pilar*), partly paid for with monies advanced him by Arnold Gingrich, the fiddle-fancying founding editor of *Esquire* magazine, who was aboard her inaugural trip, though later he decided the novelist's grossly competitive behaviour made him 'a very poor sport'. Despite refusing to accept evidence that the big ones were all females, Hemingway became increasingly fixated with blue marlin, decrying lesser quarry – he wrote to Gingrich about 'chicken shit sailfish', and when Dos Passos was taking too long to reel in a 'cuda, Hem dispatched it with a bullet. He especially detested sharks, carving his initials into their heads with bursts from a Thompson sub-machine-gun (notoriously accident prone, he once shot himself in the leg with the ricochet from a .22 pistol). He had minimal interest in fly-fishing, or the small game species of the reefs. This was angling as a branch of prizefighting. He liked the grand scale, though his younger brother, 'The Baron', was proprietor of the *Little Bimini News*, which proudly boasted being the 'smallest newspaper in the world; takes two editions to wrap a bonefish'. I do love that as a strapline.

I have never thought it necessary to like the character of an artist to appraise his or her work, but it has become a modern vogue to denigrate Hemingway's prose because of his swaggering male attitudes. Originally conceived as part of an unfinished trilogy (*The Sea Book*), his celebrated novella *The Old Man and the Sea*, known to certain weary teachers as 'The OAP and the Ocean', is indeed a fine evocation of dedication and loss, though he was contemptuous of critics who read too much symbolism into the narrative. He was so unimpressed by the 1958 film adaptation that he walked out of the screening after fifteen minutes. Rather more underrated is his unfinished *Islands in the Stream* (written around the same time, but posthumously published in 1970), which has a moving description of young David Hudson losing a rare broadbill 'grander' that disappears from view like 'a huge dark purple bird'. The book also contains a memorable description of a lady who had 'the morals of a vacuum cleaner'. Papa found Cuba inspirational, and he donated his Nobel medal to the Cuban people. He left the island for the last time in 1960, the year he handed his annual Marlin Cup to the thirty-two-year-old prime minister – the one recorded time he and Fidel met. 'I am only a novice at fishing,' dissembled the Teflon dictator (in fact, like the Argentinian medico turned matinée idol Ernesto 'Che' Guevara, he was already a keen angler). 'Well, you're a lucky novice,' replied Papa: the victor had boated five billfish for a total weight of 286 pounds, and it was rumoured he had cheated by smuggling aboard a professional accomplice to help maximise his catch (the very notion is of course unthinkable).

Ava Gardner used occasionally to fish with Papa on the *Pilar*, and he once explained to her, 'I spend a hell of a lot of time killing animals and fish so that I won't kill myself.' But eventually, like his father before him, he did just that – in effect, his final trophy was himself.

At his villa, the Finca Vigía, there were fifty-seven cats. During 1942, Papa liked to indulge in protracted hunts for German H-boats around certain habitual marlin grounds off Cuba's northern coast, an area called Cayo Romano, where he was convinced panthers roamed (survivors of a shipwrecked circus) and there was his preferred local eatery, Gato Negro, which served tasty flamingo breast, plus manatee paella. He was frustrated the *Pilar* couldn't accommodate the twin fifty-calibre machine-guns to which he aspired, but the supplies of wartime fuel came in handy. Similarly frustrated by his long absences pursuing 'Operation Friendless', his then wife, the brilliant journalist Martha Gellhorn, had all his toms castrated. (She used to refer to Ernesto's male member as 'Mr Scrooby' – well, at least it wasn't Scrod.)

In recent times, Cayo Romano has become a destination for the increasingly avid flats fishing fraternity, but when in 2011 I visited the township of Brasil, its central features were a decommissioned sugar refinery and several apartment blocks of Bulgarian vernacular-style concrete, slapdashed with primary colours. Tourists were still so scarce that you would be stared at in the street. Rice was drying along the verge of the country track to nearby Esmeralda. In the La Casona guesthouse there was that habitual tension between instinctive Cuban hospitality and the dead hand of a state tourist board. Certainly we found no manatee on the menu. The showerhead gave out electric shocks but no water, and if there was something on your pillow it sure wasn't going to be a complimentary chocolate. By the payphone I encountered a chap with a hawk on his shoulder. Government snitches lurked obviously from behind pillars. Every evening, as we chain-drank authentic mojitos at a dollar cash a pop, an old lady – conceivably some emissary from Senõr Martini – would loom from the shadows and offer our little party ripe mangoes from her bucket.

Our minibus transfer to the fishing grounds – 400 square miles of Jardines del Rey back country (subsequently devastated by hurricanes) – involved passing through a military checkpoint staffed by teenaged conscripts, who inspected our quadruplicate paperwork each morning with the gravitas of minor officials everywhere. But then we arrived at a scraggy rattan shack, and mangrove vistas that laced the salt wind with their intoxicating, medicinal tang. The permit opportunities here were exceptional, and one morning guide Eduardo and I came across a little circus of tailing *palometa*. At once I began double-hauling like a campanologist and let fly my olive Captain Crabby, upon which, with a flourish of sand, they streaked away. *'Bomba atómica,'* chuckled the surprisingly mellow guide. Around noon, where a tidal rivulet created a nice inflow, two decent specimens were finning – a sight to have Mr Crabtree dashing the dottle from his briar. 'Calma, Dayvee.' I slid over the side. The first fish bust straight off the flat, but the second continued sidling around in the current and then decided to follow one retrieve until my fly-line was inside the tip ring and I was crouched agonisingly low like a constipated anchorite. Only then did it deign to take: I turned the rod over so the knot would be flattened against the blank as my fish pelted away, but the bay was clear of snags and after twenty minutes Eduardo was able to tail out a 12-pounder. I had three further permit 'eats' that day, but blew the others. I still reckoned I was well ahead of the game.

That evening, the village schoolmaster misguidedly felt some ill-rehearsed amateur cabaret might express his community's sense of welcome. The Beatles medley proceeded sonorously enough (at any Communist *ceilidh* I am always ready with a few verses of 'Back in the USSR'), with the guides and their extended families enthusiastically singing – more so, in fact, than the anglers, because Cubans thrive on making their own entertainment – but then I was press-ganged into playing the patsy in a conjuring trick that involved a gigantic foam phallus and a sullen danseuse. I believe, though, that I counter-culturally impressed the locals with my punk-era pogo-stick interpretative dance, while singing the 1979 lyrics to 'Necro Nancy', which I had written expressly for a short-lived band the Rancid Vicars (this particular, never recorded, love song concluding, 'Take me to your abattoir!'). Through the unglazed window, Our Lady of the Mangoes appeared visibly moved.

At dark thirty next morning, I regretted our night spent drinking like horses when guide Raffa hurtled his skiff 50 kilometres through the gloom for a (fruitless) dawn tarpon session at an area known inexplicably as the Cojones de Don Kiko, and we had to breakfast off sparkling *refresco* and cold tinned frankfurters.

I decided my two sons should have a chance to see – before it dissolved – the Cuba that so beguiled me. Neither had fished the flats before. James I took to Cayo Romano, but we had poor weather. One morning, as we were wading the crème brûlée substrate of Playa Judio (Jew Beach seems a strange topographical name for this remote locale), the skies suddenly tightened, cold rain whipped across the flat and lightning played close by. We laid down our rods and huddled in the sparse mangroves until the front passed, when sun splintered down again into the shallows and bonefish were suddenly scattering about as bones on a *babalayo*'s mat. James soon latched into his first one ever, and proceeded easily to outfish me. 'Don't you want one?' he taunted, as yet again his fly-line streaked out. The old man evidently is an old man. *Coño.*

His brother Tom and I headed for the Isle of Youth in 2014, on the yacht *Georgiana*: these live-aboards are becoming more popular, but I fear they will kill the original turtle that laid the golden egg, because the area is now being overworked. I had fished much of that region before, staying at the Sol Club, a Cancun-style resort full of Canadian snowbirds, assorted beach bunnies and European poseurs wearing zucchini-tight swimwear and watches the size of Verona. There was a nearby pig farm where the swine were encouraged to thrive by listening to heavy metal music (imagine being a penned subtropical hog and also having to endure Judas Priest). Staying offshore, and therefore closer to the target zones for various species, has considerable advantages, especially at the key fleeting low-light level phases of each day. We were primarily after tarpon along the reef, and enjoyed several personal Missile Crises with rocketing specimens, most of which yawned like grizzlies about to attack, before autographing the evening air with our disconnected fly-lines. Tom boated his first 'poon after a run of twenty-odd mishaps, and was accorded the rare honour of driving the skiff back to the mothership in a rooster tail of triumph.

On my last trip to Cuba, I went back to the venerable *Tortuga*. Much had changed over the years, especially with increased knowledge of where certain species could be found. A few guides were specialising in permit, and the tarpon hunts were no longer such hit-and-miss operations. One minor tactic (which at first seemed counterintuitive) was to approach the smaller resident fish that colonise the mangrove shade by day, gun the outboard and throw streamers at them when they come out to investigate the disturbance. Ted Hughes relished laying siege to these specimens: on a trip here in March of the year he died (1998) he wrote appreciatively in the *Tortuga*'s visitors' book about tarpon 'sloping like wolves through the green light under the mangroves'.

It was to be Pedro's final guiding season, and I had him to myself. His voluble invective remained undiminished, but after two decades we knew the routine – anchored for lunch break on a strand teetering with sandpipers, language barrier intact, we sat grinning and nodding at one another like a couple of sunstruck pirates. Then, towards close of play (we had boated a dozen bones, plus a bonus tarpon) came an unprecedented utterance from the kicker bridge: 'Dayvee – perfick carsst.' In this frazzled, quasi-magical island nexus the renowned Caliban had undergone a change of heart. Prospero here nearly dropped his Temple Fork wand. 'Is mojito time,' concluded my trusty guide, and so it proved.

Dreamily tippling on the foredeck that night under a fish-scale moon, my Cohiba point glowing like a *Megalops* eye, I resolved that if angling is indeed a game of three halves – anticipation, enjoyment and recollection – then were I ever reduced to that proverbial One Week Left to spend anywhere, it would be here. The Jardines may not be quite the idyll they were, but then few places are. This refuge has meant the world to me, in its time.

I was in El Ahibe, tucking into its fabled *pollo a la naranja*, when a woman with pomegranate lips and a dress as skintight as a fisherman's finger guard invited me to escort her to the local *discoteca*. I gulped like an elderly, bronzed cyprinid. 'Angelita,' I replied regretfully: 'You really wouldn't want to mess with Changó on the dancefloor.'

17

Of Weather and Superstition

'There is a superstition in avoiding superstition ...'

Francis Bacon, *The Essays*, number XVII (1612 edition)

Anglers are frequently in thrall to the weather, and we like to cast aspersions on the adverse 'conditions' when things don't come together nicely enough. 'There's aye something wrong when I'm fishing here,' runs one Scottish ditty. Like sailors, fishermen need to become instinctive diagnosticians, developing a sense for when natural circumstances feel auspicious, including wind, wave and atmospheric pressure and, who knows, the exchange rate of the Bhutanese ngultrum. I admire the Jamaican proverb: 'Ebry day good fer fishin' – but not ebry day good fer catchin' fish.'

River aficionados tend to become preoccupied by water

height, oxygen content, clarity, character and flow; when it comes to migratory species, precipitation levels often dictate whether or not your quarry is even there for you to address. There may indeed be more to fishing than the mere business of catching fish, but that's still the main object of the exercise and a perilous drought or, conversely, torrents of turnip-hued potage (ideal for next week's visitors, of course) are liable to leave you feeling cast down. Sunshine is another variable: when you want to stalk a fish by sight it can be a boon, but in the August doldrums on a spate river any glare becomes a curse. Yet again, on a snotty March morning your prospects may be improved if sunlight begins briefly to butter up those sullen streams. Weather and fishing go hand in hand up the Start-Rite sandal road to happiness.

Many sportsmen dislike an easterly wind; some prefer a serene calm – clearly, it depends on the nature of your business. Immobile in a Hebridean loch boat without even the faintest of breezes puffing like a cherubic zephyr in the corner of an antique map, my dapping floss drooping flaccid from the rod tip, how often have I longed to conjure from my tackle bag an oxhide wallet such as Aeolus gave Odysseus, containing the winds of the world, from the Arashi of Japan to the Argentinian Zonda. Indeed, there is much sorcery associated with Nature's exhalations around the world, involving mimetic magic (whistling, or whirling a 'bull-roarer' to summon the spirits of the air) or else ceremonies to allay storms and preserve safe passage. The aboriginal Kariera tribe ensured they had flat conditions for spearing fish by attaching mother-of-pearl pendants to their genitalia and offering their deities a conch shell of human blood – which might be worth a try when a pesky gale on the Junction pool is hurling your fly-line about like a lawn-strimmer. Most folks merely accept the vagaries of weather: 'I will go tomorrow,' said the king in one Gaelic proverb. 'You will wait for me,' replied the wind.

'If you don't like the weather, just wait a few minutes' – this was probably an old chestnut already, when Mark Twain said it about New England in 1876. Weather can be extremely localised and volatile, as many islanders will attest (those notorious 'all four seasons in one day'), and remains full of puzzles – as mutable and unpredictable as angling itself. We have always been at the mercy of it, and for centuries it was by no means a trivial, chit-chat subject. Early angling books abound with 'prognosticks' about the rumbling of canine guts or teeming of pismires, and presumably the Dame (in 1496) was repeating long-embedded lore as she sensibly advised against venturing forth 'when hyt ys snowyt reynet or hay-leth thonderyt or lightneth'. Kingfisher skins were hung up to predict wind direction, jars of loaches were scrutinised for changes in behaviour, and an elaborate leech barometer was devised in 1851 by the aptronymic Dr Merryweather of Whitby. Modern science has introduced computerised wristwatches that monitor conditions for anglers, but I prefer the older ways of guesswork and vulgar errors. My favourite weatherism, scribbled on the wall of a Perthshire bothy, ran, 'Red sky at night means your sheep are on fire.'

Fish, like anglers, are moody creatures and even a basic

appreciation of biometeorology will suggest that factors like humidity, atmospheric pressure and certain winds (the scirocco, the Maltese *Xlokk*) subtly influence us all, affecting our circadian rhythms, making us torpid, irritable or depressed by turns. Most fish dislike unsettled conditions, to which they are delicately sensitised along their lateral lines. The majority are cold-blooded, so must adapt to changes in temperature (exceptions include tuna, swordfish and the opah, which resembles a gargantuan M&M); those with swim-bladders adjust their depth as the glass falls; most fish have no eyelids, so avoid glare (salmonids in particular are what one French pal described as *lumifures*). We all experience those inscrutable times when they simply won't bite, and nobody has quite worked out exactly why. Perhaps it's just as well.

In the end, angling success comes down to instinct as much as science, and a fisherman must learn to work with the grain of Nature and develop his cumulative impulses. Some people seem to have a sixth sense when they are on the water, an occasional presentiment, a singular hunch, nervous frisson, subconscious inkling or feeling in their water that something is about to happen – call it what you will. Ransome dubbed it 'the benign moment', and compared it with those freak silences that sometimes fall on an assembled company (at my boarding school this was referred to as 'an angel passing over'); he records that in czarist Russia folk would say, 'A policeman is being born.' Professor Luce explains this phenomenon as 'the principle of psychological sympathy', whereby the angler is responding to a combination of atmospheric nuances, and

Falkus astutely ascribes 'The Feeling' to 'part of a shrivelled instinct-to-hunt'. Something in the ether seems freshly charged and tense – it may just be that the air is newly ion-ised after a downpour, or nothing as straightforward as that. It could be that your immediate surroundings simply smell slightly different. You become unusually attuned and atten-tive, and there's an epiphanic sense of conspiring. You may even experience a type of predictive *déjà vu* – what the blind poet Borges called 'a Remembrance of Things to Come' – and right then you are living deep in the fishological moment.

In many cultures there has been a widely held fantasy that some individuals can control or modify the weather, which harks back to the days when the very survival of a tribe might depend upon it. In arid regions, the rainmaker appeared weirdly powerful, usually a shaman or hydromancer with his rod-like wand and enchanted hair that supposedly cho-reographed the winds and water (Prospero belongs to such a tradition) – and in places such magical ceremonies persist. The Maasai still burn cordia wood and chant *'Ai, tasha'* ('Rain, fall!'), and the Australian Yeidji headsman mixes his hair with iguana entrails and abandons his wife for two days (an unchallengeable bunking off being one badge of patriarchal societies). 'Send her down, Hughie,' implore contemporary outback dwellers, pleading with their more informal weather deity. Physical rain charms are often pisciform, and freak showers sometimes actually contain fish (in 1931, a rain of perch halted the traffic in New York). In his spell-like Laureate poem 'Rain-Charm for the Duchy' (1984), Ted Hughes heralded the breaking of a long drought in the West Country rivers with a fertility chant that sees the waiting salmon respond to the onrush of sky water: 'those other, different lightnings, the patient,/ thirsting ones'.

Hughes's verse references a widespread and lively history of belief in thunder magic, which seeks to explore how, in the hands of certain sky gods, the weather has become weaponised. The Babylonian Adad with his lightning spears, Buddha and his thunderbolts, the vengeful Jahweh scattering his enemies, and redheaded Thor in his goat-drawn chariot were all thought to express their displeasure by sending fire from heaven. Thor's magic hammer Mjollner struck sparks from the dull skies, and splinters could be turned up by ploughshares – thunderstones and elf shots that were treasured as heirlooms and talismans (in reality, shards of meteorite, prehistoric arrowheads, or even fragments of fossilised cuttlefish). Finnish anglers used to rub them along their rods for good luck. Around the world there are various lightning birds: the Zulu and Xhosa have their impundulu – sometimes identified with the hammerkop, a wader with an electric-looking crest – which creates thunder through its wingbeats, and (like Changó) strips the bark off trees. Kalahari bushmen believe you will be lanced by lightning if you interfere with its nest. Native Americans revered similar figures – Emily Dickinson's lightning storm 'showed a yellow beak' – and Thunderbird is the given name of that potent, fortified wine that has fuelled many an American hobo.

Lightning is usually crooked, may strike upwards and can arrive out of clear skies up to 10 miles distant from any storm (that 'bolt from the blue'). It can appear yellowish or pink: fishing for Nile perch on Lake Victoria in 1983, Ted Hughes witnessed a blue and green display, 'great skyfulls of blazing thorns' (letter to Barrie Cooke, 23 October). You can survive its strike (golfers seem particularly prone), though it moves at over 60 miles per second and delivers a hundred thousand gigawatts of energy. Occasionally it will strike twice in the same place, like a fish. For centuries prior to Quaker polymath Benjamin Franklin's development of the conductor, it was thought that ringing church bells would ward off celestial electricity, despite the fact that spires themselves make prime targets. Some mediaeval bells were even inscribed with *'Fulgura frango'* (I break up lightning), and the safety of village campanologists was not enhanced by the practice of storing the local reserves of gunpowder in steeples. Waving your hydromantic graphite wand around while standing up in a vast expanse of open water is a similarly ill-starred idea: several times I have had to lie prone in a flats skiff to avoid being the highest point for miles around. Captain Ahab (whose nickname was Old Thunder, and who tempered his harpoons in blood and lightning) actually becomes a human lightning conductor in one transfiguring scene.

In ancient Rome the prestigious college of augurs interpreted the mood of certain gods by studying the behaviour of birds, shooting stars and lightning (divination by the latter being ceraunoscopy), and great store was set by their pronouncements on everything from military stratagem to the fate of individual families. The laurel, being sacred to Apollo, was thought to be immune to lightning, so during thundery weather nobles used to wear laurel-leaf chaplets (Emperor Tiberius especially dreaded such tempests; the brief reign of Emperor Carus was cut short in AD 283 when he was killed by lightning on his campaign against the Sasanians). In England, our venerable oak was deemed to possess identical qualities: that wooden acorn on the window-blind cord of older country houses is a vestige of such belief. 'What is the cause of thunder?' a maddened King Lear enquires of the mock-mad Edgar during the storm (offstage, it would have been caused by cannon balls being trundled down between wooden slats). The notion endures that if you could somehow translate the thunder's language you would be able to understand the voice of the President of the Immortals (T. S. Eliot closes his panoramic *The Waste Land* with allusions to the thunder prophecy from the Sanskrit *Upanishads*) and, perhaps particularly nowadays when we seem to have disfigured our natural patterns of weather, there endures a fascination with wondering just what the lightning might be spelling out momentarily across the sky with its scrawl of elemental calligraphy.

Considering its earnest, multiple uncertainties, there can be little surprise that angling is a pursuit cross-hatched and shot through with superstition. Although it literally means 'standing still in awe', superstition is something that convinces other people of its valency, but not you – it is thus often shrugged away as so much phonus-balonus. But I have always thought there was something beautiful about this mongrel offspring of coincidence and dreaming: it is calcified piety, the stubborn belief that remains when science and rationale have sluiced away most of the nonsense. It is what allows you some comfort in our Cloud of Unknowing. I agree, you could try religion – but I believe the gods were invented by people only when they felt their own magic was not really working.

In fishing, as in life, there are two types of calamity: misfortune to yourself, and good luck to others. Let's be honest about this. I don't want to sound like some ghastly motivationist, but you need to work at maximising your own luck, and one way is to cultivate positivity – 'When Luck enters, give him a seat!' runs the excellent Yiddish maxim. Another gambit is to spend as much time as you can on the water, because luck interplays with skill ('You know, it's funny,' said Gary Player, 'but the more I practise the luckier I become'), though beginner's luck often seems unaccountable. Unfair as it is, some folks are just plain fortunate. An Arab proverb runs, 'Throw a lucky man into the sea, and he will come up with a fish in his mouth' (I've heard some bitterer versions involving a cesspit and a golden fob watch), but for us ordinary mortals – assuming you haven't acquired a guardian angling angel – there's an argument for supplementing your efforts with a little alternative help. In days of yore there were grimoires like *The Book of Secrets* by Albertus Magnus (1502) that included actual spells and incantations for attracting fish, but these probably worked as effectively as modern-day dating guides. Walton's was not the last era to lend credence to alchemically infallible concoctions of near-necromantic complexity, such as those bait additives requiring human 'mummy' (not the Egyptian type, but merely human fat which was among the builder's perks of a common hangman). My nostrum is the lucky charm, which fends off disasters from skunking to shipwrecks (with no further comment concerning helicopters). After my second grilse of the morning landed out of one pool, a gillie on the Mourne summed up my position: 'Sure, ye have the luck of a fat priest.'

From steeplejacks to gladiators, all the precarious professions have their traditional superstitions and accompanying lucky mascots. They help us feel confident and strong in the face of uncertainty. Alpinists used to sew inside their lapels an eagle's tongue (the lightning bird that is always safe aloft), gamblers shun the four of clubs (or Devil's Bedstead), astronauts avoid black, bullfighters make a fetish of their suit of lights. Actors are notorious superstitionists: I became a little jumpy when I mislaid my theatrical mother's rabbit's foot (many burrowing animals were deemed close to netherworldly spirits), and my naval grandfather wore her birth caul in a locket around his neck to avert drowning (a Blitz bomb finished him off in the end). My father, in his amateur jockey days, always carried a small, engraved golden horseshoe, which I in turn passed on to my son when he was posted to Afghanistan. But I suppose such touch pieces depend largely on what you desire of them: at the height of Madonna's pop stardom there was a rage for wearing crucifixes, and one fan enquired at her high-street jeweller whether they had an example 'with one of them little men on it?'

Writers can be a ritualistic lot. Some whisky distilleries never brush away the cobwebs around their stills for fear they are an indefinably subtle part of the magic, so similarly, if peculiar observances seem to work for you, why take the risk? Samuel Johnson always entered a building right-foot first. Voltaire used the back of his naked mistress as a writing desk – an early instance of the laptop. Ernest Hemingway wrote standing up, often in kudu-skin loafers, and was profoundly superstitious (thus all those Cuban cats), but, like many authors, he refused to discuss it for fear of diminishing the effect of his beliefs. Papa was forever acquiring corks, nuts and shells as amulets; once he owned a hat box containing a shrunken human head, which required some explanation to the local traffic police when his Lancia was involved in a Spanish prang.

Such adherence to charms that stave off adversity and disappointment may seem ridiculous to some, but they have become an integral part of my life, both on and off the water. My desk supports an alignment of necessary objects which are moved but once a month for dusting – any other disturbance is a hostage to Fortune. I have managed to write several books with their apparent complicity, and don't wish to adulterate any positive influence. These tokens include a brick portion from the novelist Faulkner's garage wall, a Mexican travelling shrine triptych, a miniature golden creel, a statue from Easter Island, an unopened bottle of Special Forces pepper sauce, a polar bear claw, a rhea egg shell painted with lightning shapes by my son, a soapstone Alaskan halibut, a Ch'i knife coin my father acquired from the Forbidden City in the 1930s, a packet of Shark brand razor blades, and a cocktail mat from the Upland Goose Hotel in Port Stanley. Writing, like fishing, is often an act of faith; but, as an angler, I don't just believe in good luck – I need to believe in freaking miracles.

Perhaps no occupation has been more freighted with superstitious lore than that of the mariner and commercial fisherman, where sheer survival at sea is interwoven with the need for bounty so the community back on land will prosper. The champagne bottle gashing the bows of a newly launched vessel may have replaced earlier blood sacrifices, and I doubt many Whitby trawlermen still wear a sheep's bone known as Thor's Hammer around their necks to stave off storms, but there are older seafarers who still believe certain seagulls are reincarnations of the souls of drowned men (the practice of wearing a gold earring may originate from the fear that a washed-up corpse would be denied a proper burial unless there was something of value to fund it). Onboard taboos used to be numerous: clergymen, red-heads or women were widely avoided as passengers (John Aubrey specified that having a 'whore' on a boat 'does cause a storme'). Among the fishermen along the eastern seaboard of Scotland, if you so much as encountered any of the above while en route to your mooring you were to swear a mighty oath, 'May the divell tear oot yer tongue!', followed up with an optional sock to the jaw. There abounded verbal taboos:

a seal was 'bald head', and salmon variously 'laddie', 'the cold man' or 'the gentleman'. Anglers have inherited some of this circumlocutionary tradition – I once heard a splash behind me as a Hebridean gillie was rowing us back up a loch, and asked, 'Was that a fish?', to which he replied, 'No, it was a sea trout.'

After a brisk morning's business on the bonefish flats, I was unpacking my picnic box when my Caribbean guide spotted the banana. 'Don't even say the name of that fruit, man,' he warned, 'it bring real baaad luck.' Too late. As we ate, oily cumulonimbus clouds rolled up like miasmas from Tolkien's Mordor, and we were soon fleeing an electrical tempest at full throttle. This belief is strangely common, and seems to hail from Hawaii. It may be because slippery skins truly are perilous things to have around a deck (one has heard of fatal accidents), but I think it's more because the phallic banana is a prime candidate for the original Forbidden Fruit (its leaves are even handy for making improvised clothing, as Castro recognised during his *Período especial*). Taboos are usually rooted in common sense, but over time some practices – concerning diet or sexual behaviour, say – might become inexplicable or absurd to outsiders. *Tapu* comes from Polynesian culture, where rules to be observed at sea were rigorous (anything ill-omened while fishing was called *puhore*), as were the elaborate rituals in preparing nets and lines, which women were never allowed to touch – the menfolk would urinate on their tackle if it was ever polluted by female hands, a habit that persisted among the Maoris within living memory.

Leisure anglers don't care if civilians find their private rituals risible, but these tend to be eccentricities rather than a matter of life or death. An unshakeable faith in lucky rods – some of them individually christened, like trusty battle swords of old – is an understandable foible, and in the Highlands it used to be held that wood from a mountain ash was ideal, since the rowan's fruit is a fortunate red, and similar to the spots on wild trout (its timber was thought auspicious for boat building also, and lucky for the 'rantree' crossbeams of your house). One Gaelic proverb translates as, 'Rugged tackle, a stolen hook, and a crooked wicken rod', reflecting perhaps the native Scot's enduring credo that after nightfall all wildlife becomes common property. Favoured floats, hooks and flies are common to many fishers, as are certain items of clothing. I have a particular stratum of superstitions about hats. As the Quangle Wangle's milliner would doubtless attest, there can exist an intimate connection between headgear and mindset. Gone are the days when anglers sported high-crowned sugar-loaf hats, Walton style, or beaver tricorns, or silken toppers with the gut cast wound around, but the choice of titfer for a particular expedition seems to me a matter of singular importance. My green felt trilby, a veteran whose underbrim is blotched and etiolated by a decade of bug spray and the sweat of my summer brow, is matched to sallying forth towards chalk stream or the smaller salmon waters, whereas any one of my seventy-odd baseball caps (also known as 'gimme' caps, as in 'hey, bud, gimme one of them hats') will rather see action on lakes or at sea. I occasionally swap hats during the day if I am not faring well, so I must admit my superstitious streak, though possibly harmless, may have gone beyond quirky.

I feel under-protected against calamity, too, if I do not tote a number of accumulated amulets to exude prophylactic magic that averts misfortune and simultaneously to radiate positive magic that works wonders and invokes the benevolence of the fish gods: my *sine qua non* selection for any foray includes a bead and creel charm made for me two decades ago by my daughter Laura, a hook-fastened braided cord bracelet (twisting threads famously confounds your enemies) and a treasured Maori *Hei Matau* fish-hook pendant. There is an astonishing collection of these in Wellington's Museum of New Zealand, and they comprise a cultural symbol of great complexity right across Oceania, being associated with the Creation myth of the trickster deity Maui. Using such a fish hook fashioned from the jawbone of his heroic grandmother Murirangawhenua, a line made of his father's twisted sinews (Tiritirikimatangi, which never took off as a tippet brand name), and for bait his own earlobe, Maui towed up a vast land mass behind his canoe, which gradually broke away to form the Southern Islands (including New Zealand). He was slain in Hawaii and his blood forever tinges crustacea pink. Although my charm is of jade, Maoris were keen on hooks made from human bone and sometimes mounted the shrunken heads of their enemies (*Tama-a-hara*) on the gunwales as they trolled kiwi-feathered lures for tuna and 'cuda, so the skulls shook as fish pulled at the line. I'd like to try this sometime for ferox trout on one of the deeper Scottish lochs.

The drawback to a reliance on charms is that you can become too dependent on them. For several years after a trip to Alaska with my two sons, I faithfully carried around a souvenir coin imprinted with a chinook salmon and inscribed 'Determination' – it slotted into the ticket pocket of my chinos like a miniature disc drive, and I had toted it from Laxá í Kjós to the Tongariro, apparently with productive effect. In December 2019, while attending a conference at Pembroke College, Cambridge, my touch piece disconcertingly and mysteriously went astray as I was changing for the dinner at which I was due to deliver a speech about Ted Hughes and fishing. I felt stricken and bereft.

Pembroke is Ted's old college, and that agreeable literary scholar Mark Wormald had previously shown me many of the manuscripts and artefacts he had acquired for their Hughesiana collection – including some of the Laureate's old tackle. I knew Ted was a passionate believer in omens, luck, shamanism and synchronicity, and was intrigued to see the swivels on his spinning lures which bristled with nylon stubs where he had refused to cut through the knots (another ancient superstition to avoid the release of magic). After my little speech, quite out of the blue Ted's fishing friend Ian Cook presented me with an envelope which contained one of the poet's own Blair spoons, its swivel eye festooned with several such 'whiskers' – a new talisman to replace what had unaccountably disappeared. Later that night, in my college room, I looked at my wristwatch and it was twelve minutes past twelve on the twelfth day of the twelfth month, and on the windowsill my trophy glowed in the pale light of the last full moon of the decade, regarded as a time of propitious spiritual activity. I felt it had been more than a fair swap. I stood still, in awe.

It surely is the magical aspect of water that generates much of the angler's transport of delight, as well as chronic perplexity. A source of both bounty and destruction, water has been revered in practically all cultures – cleansing, sustaining, potent, potable, portable and formidable, it enjoys sacramental associations of baptism and absolution, as well as sustaining many fluvial and deluvian myths. If you held hands over water, a bargain was considered binding; it was thus that Robert Burns 'plighted his troth with Highland Mary'. Individual streams were personified with their own spirits, often female – Verbeia for the Wharfe, the Tamar's Tamara – and certain rivers had to be propitiated with annual sacrifices or accidental drownings ('bleedthirsty Dee needs three', whereas 'Bonny Don needs none'). Hydrolatrous rituals have been a common feature of the cycle of the seasons, whether it be the blessing of nets, the placation of the god Shony by pouring a quaich of whisky into various Highland waters on opening day of the salmon season, or the ceremony of the Wonkgongaru aborigines (who regard themselves as being descended from fish), wherein, to appease their divine water bringer, a chieftain is constrained to coat himself in red ochre and pierce his scrotum with fishbones. (In feverish dreams I sometimes feel I saw this enacted in the gillies' dormitory on a Hebridean estate.)

The seasonal behaviour of fish used to be deemed semi-miraculous and was scrupulously celebrated. Many North American tribes treated the migratory arrival of the first salmon or sturgeon with great totemic ceremony, and the Yuchi performed a rippling dance to encourage their progress upstream, waving their arms like fins. It may also be that in the oft-maligned Morris dance (originally a mediaeval routine that was considered vaguely Moorish), the coloured ribbons signified the welcome arrival of estuarial mullet, with their yellow and pinkish hues. There was also once a ritual Celtic 'trout dance' that mimicked the rise forms of fish responding to the spring hatches of insects, and a salmonid seems to have slipped into the gene pool when the spirit of Irish letters was being formed. Yeats, who was a keen angler and described in his elegant autobiographical poem 'The Fisherman' that distinctive 'downturn of his wrist/ when the flies drop in the stream', tells in 'The Song of Wandering Aengus' of how, on a hazel switch baited with a berry, the poet's insomniac alter ego catches a trout that turns into a water sprite and he spends the rest of his days in her pursuit (a beguiling and spare analogy for the angler's unrelenting quest). The realm of faery and the world of water are closely commingled, too, in various versions of the Salmon of Knowledge (*eo-fiosach*), an aetiological myth that accounts for the coming of poetry and art. The salmon (*eo* sounding marvellously close to hero) is a motif of good health, an avatar of the underworldly gods and intimately associated with the migration of souls. At the source of the Boyne near Tipperary lay a well known as Connla's Fountain, overhung by nine hazel trees, the nuts that dropped from which were swallowed by the enchanted salmon. From a pool named Fec's, the fair hero Fionn mac Cumhaill caught and consumed a salmon which endowed him with the power to foretell the future. In one early version by the bard Amergin, the *eofis* sings, 'I am a lure from Paradise.'

All around the Celtic world there were sacred wells, scraws and *fuáran* that in their thousands were places of pilgrimage, prophecy, cursing and healing. Worldwide, there is a basic fascination with peering down into wells and seeing the distant reflection of your head as if on a coin, and all have been valued as sources of sweet, mineral or spa-like water. This has been gratefully acknowledged since pagan times by votive offerings (including coins, shells and pins) and seasonal dressing – in England with floral tributes, though these were hard to come by further north and west, so ribbons, 'clooties', rags and tippets were substituted by celebrants and petitioners. Their curative properties were sought for treatment of multiple afflictions, from epilepsy and the plague to tooth worm, cholic, infertility, bladder gravel and outright insanity. These natural aquatic powers were sometimes boosted by dropping in the heads of slain Norsemen, a practice not encouraged by the early Church which, in characteristic fashion, duly annexed many such sites to Christian hagiography. A large number of them disappeared during the Agricultural Revolution, but a few still exist, and their waters were being worshipped up until the Great War. 'Cranfield pebbles', the amber-like crystals (probably gypsum) from St Colman's at Lough Neagh, are still collected as charms against drowning.

Like certain temple lakes of the Orient, where tame cyprinids can be summoned by bells, these wells often held 'holie fishes', most commonly sacred trout, which symbolised the mystique and purity of such places – though, I imagine, they may also have been introduced to keep down the levels of insect larvae. In Banffshire's St Michael's Well there was said to be a prophetic fly, the surface movements of which might indicate your prospects of recovery. At St Ciaran's (County Meath), there were spectral trout you could only glimpse on the first Sunday in August, while at Tobernalt (Sligo), a blind man had his sight restored by a sudden piscimiraculous vision of 'a speckled trout with a white cross on its back'. Some of these guardian spirits were immensely long lived, and at Kilmore there even swam two immortal fish. In times of famine, certain wells miraculously yielded salmon, and the Cornish St Neot had a personal stock continuously replenished by an angel (you just can't get the staff, these days). The identification of fish with the marvellous is an essential part of human history.

At one time there were more than six hundred holy wells in Scotland alone, many of them in the islands: on Eriskay and Barra, fishermen took healthful draughts before putting to sea, and even in Victorian times the Stornoway doctors would take a bottle of preferred well water to patients on their deathbeds. On remotest St Kilda were several wells, one of which, Tobar nam Buadh (the well of virtues), was so renowned as a diuretic source that folk sailed there from Harris, a perilous 40-mile sea passage. I never did find it when I visited (there was said to be a trout in that one, too), but nor was there ever a saint called Kilda: that beleaguered island group is possibly named for *kelda*, the Norse name for 'well'.

A dozen Scottish water sites were dedicated to the composite figure of St Bride, who has become conflated with the Celtic goddess Brigid. Historically she was a charitable Irish nun named Brigit, baptised by St Patrick himself, and famed for her healing of lepers. She died in 525 on 1 February, which tallied with Imbolc, the pagan festival that marked the beginning of spring, and a traditional day for weather prognostication. Her legends were later embellished so she somehow became identified as the foster mother of the Christ child, 'Mary of the Gael', a woman so delicate she could hang her gown upon a sunbeam. Sacred to her memory was the oystercatcher, who protected Jesus when he was pursued by his enemies along the Hebridean shoreline, and ever after those birds have carried the marks of the cross upon their backs. St Bride is widely commemorated, not least by the spot in Fleet Street where a church has stood since Saxon times. Down south she is the patron saint of the printing press, but on the wilder shores there are those who still pray to her for good weather as they venture out to sea in their fishing boats.

It may just be that there is something magical and uncanny about fish, which accounts for their abiding allure. Or perhaps Thomas Edison had it right when he said, 'We do not know one millionth of one percent about anything.'

18

Creatures of Light

'So we found the end of our journey./ So we stood,
alive in the river of light/ Among the creatures of
light, creatures of light.'

Ted Hughes, 'That Morning' (1980)

At four in the morning on 30 March 1867, an agreement was
signed whereby Russia sold to the United States a piece of
land twice the size of Texas for just under two cents an acre.
It boasted 35,000 miles of coastline (more than that of all the
other states put together), a string of volcanic islands, the highest peak in North America, several climates ranging across
some twenty degrees of latitude, and untold mineral wealth –
although Alaska graduated from being a mere 'territory' to
becoming the forty-ninth state only in 1959, not long after the
discovery of its oil.

Palaeo-Indians had crossed the land bridge from Siberia tens of thousands of years before the continents were separated by what became known as the Bering Strait, named after the Danish-born explorer who finally reached the Alaskan coast from Russia in 1741, then died from scurvy. At its narrowest just 53 miles, it was to become a geopolitical barrier between Russia and America, and during the Cold War was known as the 'Checkpoint Charlie of the North'. The name Alaska may derive from the native word *alaxsxaq* ('the place where the sea breaks its back'), and the first visitors to encroach on the Yup'ik people's traditional hunting grounds were fur trappers, especially those in search of valuable sea-otter skins – having no layer of blubber, its pelt is extraordinarily dense, with up to a million hairs per square inch. Thanks to slavery, hooch, murder and smallpox, some native populations were soon reduced by 80 per cent, as successive waves of whalers, missionaries and other prospectors ventured to exploit the resources of 'Seward's Icebox' (after the secretary of state who had signed the acquisition papers). For many years it was a neglected region synonymous with remoteness and the frozen wild. Just prior to the achievement of Vitus Bering, in *Gulliver's Travels* (1726), Jonathan Swift set his imaginary land of giants (Brobdingnag) off what was to become the Alaskan coast. It is perhaps no coincidence that so many things here are king-sized, including the fish.

In July 1998 I flew into Anchorage from Seattle, seated between a Hawaiian nun and a youth with turquoise hair. Both of these exotic visitors would have fitted right in with the state's only cosmopolitan city, once a shanty town for railroad workers and sourdough prospectors and now an agreeable mix of high-rise and sprawl with as colourful a cast of inhabitants as the cantina scene from *Star Wars*, and comprising a comparable hub for air traffic – Alaska has the highest per capita rate of aviation fatalities in the USA, and one in five Anchorage men holds a pilot's licence. My destination was the gateway settlement King Salmon (a place name that's right up there with Intercourse, Pennsylvania), from where we headed in a bush plane towards the Bristol Bay area, sometimes called the Salmon Factory of the World. Buddi the pilot's little Beaver hummed over a remarkably green landscape, splotched with innumerable lakelets and veined with creeks and sloughs, towards Katmai Lodge some 300 miles from the nearest road, sited on a bluff overlooking the Alagnak river (also known as the Branch, for its intricate braided side streams). It was one of the largest fishing camps in the state and renowned for its superabundance of salmon. I wondered if such reports were an exaggeration, but as we banked near the airstrip I realised I could actually see from the air shoals of fish rippling up the shallows.

They arrive every summer in their millions, overlapping runs of five different salmon members of the genus *Oncorhynchus* (the name, meaning 'hook-nosed', alludes to the kypes they develop towards spawning). These are keystone species for the nutrient systems of so many rivers in Southwest Alaska, bringing their lavish proteins from the Pacific back to the hinterlands of their birth, where they spawn in different

reaches and at various times during the autumn. The redds
are then gashed with thrashing vermillion, while char and
rainbows become gluttons for the eggs, then feast on gobbets
of disintegrating corpses, because, unlike their Atlantic cousins, all Pacifics die after breeding, their bodies entering the
food web and benefiting some fifty other species – bears, birds
and bugs – whose lives are interdependent with this extravagant annual migration. None ever lives to see its offspring. The
stench in these headwaters becomes unforgettable, as new life
and mass destruction frenetically commingle.

Rarest and mightiest of this pentarchy is the king (*O.
tshawytscha*), a thick-shouldered magnifico with flanks of
blush wine and a gunmetal mouth, also known as 'chinook' or 'tyee' (the Nootka word for 'big brother'), a real
Brobdingnagian fish. In rivers like the Kenai, it may grow
to over 100 pounds, a matinée idol, handsome, pugnacious
and insolent. When fresh run in spring, its voluptuous flesh
(which can range from off-white to sunset orange) is so fatty
you have to take care your grill doesn't catch fire, and, bled
out first to avoid any ferrous taste, the king's meat is greatly
prized in fancy urban restaurants as well as being popular,
smoked, for lox in bagels.

Sometimes called 'calico salmon' for the skin markings it
develops in fresh water, the chum (*O. keta*) has never much
been as sought-after for human consumption, its flesh tending
to be flabby. However, there is a Japanese market for the preserved eggs (*ikura*), and the chum has long been harvested for
winter dog chow, thus designated 'the gasoline of the North'.

A typical sled team was estimated to get through a thousand fish each winter, and chum were also nicknamed dog salmon, though this is probably because of their ferocious, canine dentition and frankly dogged attitude when angled for. Chums coarsen and colour up quickly on leaving the salt, and begin to look as if they have been in some Klondiker's barroom brawl. In the ocean, they like to guzzle jellyfish.

Depending on region, July usually sees the main run of *O. nerka* – red, blueback or sockeye – which are most prolific in river systems that include a lake for part of the juvenile life cycle, and there is also a non-migratory strain known as 'kokanee', or simply 'yank'. The name sockeye has no ocular relevance, but derives from an indigenous word, *suk-kegh*, meaning 'red fish'. This alludes to the lurid epigamic colours it later adopts, when the lithe ocean chrome sheen is replaced by garish green and red, so that the sockeye resembles a section of beetroot dipped head and tail in spinach purée – a fish painted by Matisse after a beaker of absinthe.

I have long felt sympathetic to the unprepossessing humpy (*O. gorbuscha*), which is the smallest and most numerous of the Pacifics, also called the pink. Running in alternate years only, that Russian binomial means 'hunchback' – and they soon become additionally afflicted with a crocodilian snout, and grow as grotesque as gargoyles. Unless caught directly off the tide they are unlovely and slimy, though still good to eat if smoked or canned. They tend to stray during migration, and in 2017 a number (maybe from the Kola Peninsula) turned up in British rivers such as the Dee and Ness, and were deemed unwelcome intruders. Many anglers consider them a pest that intercepts flies intended for glitzier or supposedly nobler quarry, such as the late-running coho, or silver (*O. kisutch*), which is glamorous and aerobatic and all things the poor pink never is. In Siberia, Korea and Japan there is also a cherry salmon (*O. masou*) – smallish, stocky fish that run the rivers at cherry blossom season. The male sports cerise bar-chart markings on its flanks as it prepares for the matrimonial deathbed. Of the runs I have witnessed, Pacific salmon migrations are surely among the Wonders of the World.

'Forget everything you may have learned about fishing for salmon back home,' is the advice I am given the first evening at Katmai, sitting on the camp deck with our group host Arthur Oglesby. As well as being a prolific author and photographer, I suppose that Arthur (then seventy-five) had probably taught more British anglers to fly-cast than anyone else, and, despite his beady-eyed charm (there was more than a touch of Cary Grant about him) and the fact that he was impish, vital and simpatico, I went in awe of him. From Norway's once-fabled Vosso river, where his friend Odd Haraldson had the lease on the Bolstadoyri water, Arthur had killed several astonishingly large salmon (*storlax*) including, in 1973, a specimen weighing 49 pounds and 10 ounces – 'I refused to call it fifty pounds,' he explained, so with characteristic panache he asked his host, 'Perhaps I could come back and try again?' King Arthur, as I called him, had been running courses on Speyside since the 1950s, and in 1978 invited Hugh Falkus as guest instructor – they made a powerful double act, until Falk caused the inevitable falling-out. The polar opposite in temperament to Hugh, Arthur's persona was studiedly modest. Despite his fame in game-fishing circles, he made no claims to being a great stylist – he confessed his punctuation was so arbitrary that one editor labelled him 'Arthur (Semicolon) Oglesby' – and he concludes our nightcap session by saying, 'The more you learn, the more you are aware of your ignorance.' For several years we were friends, and he was one of the nicest angling experts I ever knew.

I am to cut my Pacific teeth next morning on sockeyes, and indeed do have to set aside what little I know about salmon fishing. After two years away at sea, the reds return in such numbers that you can sometimes hear them flapping and shuffling their way over the gravel bars, obligingly migrating through the shallows. Unlike some rivers that are coloured by glacial silt, the Branch runs clear and you can make out the pulses of salmon swaying their way upstream, in places several abreast just beyond your rod-tip – not quite the proverbial 'three feet of water, one foot of fish', but near as damn it. They will not deviate much to intercept your lure – generally a nymph-style pattern with a ribbon-wrap of lead, plunked across and drifted down through the shoals – and since in the sea they are planktivores that sieve-feed on microscopic crustacea (they can detect essence of shrimp diluted one part in a hundred million) and never eat in fresh water, it's anyone's guess why on earth they take at all. Some pundits claim you are only ever foul-hooking them in the mouth – but although you will snag a few accidentally, if you concentrate on technique you soon develop a knack. On your single-hander, they fight and leap and are as valiant as any summertime grilse. Many throw the hook, but as guide Scott reassures me: 'Don't worry too much, Dave. There's another two million coming right up behind him.'

You do need to recalibrate your superlatives a little here, as such plenitude is clearly unfamiliar. I hooked more salmon that week than I would have managed in an entire decade back home, and the experience of spending several hours at Fin Fever Central becomes a touch unreal, like an Escher staircase. In Paradise, a guy never gets skunked. Mind you, one does have to watch out for two of Alaska's most dangerous creatures – first, the mosquito (plus a type of blackfly known as whitesox, or flying piranha, which can certainly nail you through a thick mackinaw shirt); and then there's the Alaskan grizzly, which can weigh 3,000 pounds, covers 50 yards in three seconds, loves salmon and in places is so 'food-conditioned' it will emerge from the riverine brush at the sound of an angler's reel ratchet. The bankside trails in this particular Eden are ominously dolloped with fishified, berriful ursine scat, serving to remind you this is not simply a dream, wherein it's only a matter of time before the whine of an alarm clock will yank you up into the wakeful sky.

King Arthur has landed so many reds over the years that they don't excite him much any more, but he comes with me the next day when Scott races our aluminium river sled 8 miles downstream towards tidewater for my first try at a chinook.

Cliff-nesting bank swallows loom the air for insects in the brittle morning light as we scoot past deadfalls and sweepers and the odd rough meadow with stands of magenta fireweed before we roar into the margins and begin to ready the gear. Kings like to keep their chins close to the riverbed as they run, and popular methods include back-trolling a plug from your boat (similar to Tayside harling) or drift-fishing, whereby you bounce a bead and plastic egg 'cheater rig' down behind the boat – if you're not snagging the bottom from time to time, you're doing it wrong. I have tried something similar for tiger-fish on the Zambezi, and am keen on any technique that hooks me up. Tick tick, goes the tackle – then, *zowee*, a side strike, and we are in. That didn't take more than fifteen minutes, and the Loomis bait rod is heaving against the weight of our fish: I pump and agitate, and before too long a nice king comes to net, and is dispatched for the freezer (you are allowed one per week, and he weighs 33 pounds). I make a feeble joke about it being a Wenceslas – a 'good king' – and Arthur unscrews the cap from a Black Label miniature, to mark the moment. By any standards, this is a fine, buccaneering fish, but the authentically recherché method is to hook one on the fly, and I'm assured that's what we have to try next.

Up at the head of Aquarium pool, Scott has me double-hauling a Teeny 400 sink-tip line and a vane-fly lure that looks like a bullfinch chasing a UFO. I have to mend incessantly, to achieve the depth where I can fish to the king's face. I miss the first pull, then hit a magnum jack around 8 pounds – the king equivalent to a grilse. Second time through the run,

I follow instructions and let the lure dawdle longer on the dangle. 'There you go, Dave.' The line draws off my reel and there's no genteel lifting into him – you strike a king with your entire body, and then some: o-bop-she-bam. I rear back with my 10-weight, and the carbon blank crunches off like a stick of celery, snapping just above the cork handle. Scott grabs the top of the rod and, somehow, after a frenzied ten minutes, we wrestle the fish to the boat. It's an 18-pound hen, her chrome lightly scumbled with bronze – the fish I had come for, in this empire of the *Oncorhynchus*. 'Man,' says Scott with a shake of his head, 'I never saw such a hookset.'

Later, when we're alone over sundowners, I ask Arthur about Hugh. 'Oh, David – you knew him. Dear old Falk; he was a great character. I admired him tremendously as an author – it made me want to throw my typewriter out of the window. But, well: he was a cunt.' Which was a coincidence, because one night at Cragg Cottage, a decade previously, when we had polished off his Scotch and were, perilously, working on a litre of Spanish brandy, Falkus had said to me, 'Arthur is a charming individual. But he's a cunt.' So much for halieutical history. This harsh, reciprocal verdict was not true of either man.

In his classic study *Coming into the Country* (1976) the great geological chronicler John McPhee describes Alaska as a foreign country significantly populated with Americans, and this sense of otherness – often marketed as a 'Last Frontier' experience – underscores even a visit to a comfortable, if remote, fishing camp. Visitors frequently report a feeling of liberation in such wide expanses, a glimpse of some new perspective on Nature, even fresh, visionary wisdom. Out in the unforgiving bush, it's easy enough to imagine Alaska has plenty of 'thin places' where barriers of spiral time may have collapsed, and the Northern spirits swoosh in and out – though in my experience you won't come across this sort of thing by deliberately seeking it out, even with rod in hand.

It remains a very sparsely populated state. Anchorage is home to nearly half of its inhabitants (those living in the bush like to describe the city as being only half an hour's plane ride from Alaska proper), and it is said if New York had a similar population density there would only be thirty-six people living in Manhattan. Well, at least you could then get a table at Masa. Almost everyone seems to be originally from 'Outside', as they call the Lower 48, and the territory has tended to attract its fair share of idealists, nonconformists, hermits, dropouts, fugitives and screwballs which, at least on the surface, seems to make for a persistent culture of macho, carnivore, independent-spirited jocks, river rats, bush pilots and assorted riffraffia, many of whom are impelled by wanderlust. Moose, eagles, fish, folk – migration is the default lifeway of the 49th. Back-country existence remains raw rather than bucolic or Bohemian. The weather is so gonzoid it will turn around and bite you in the ass (its violently changeable climate probably disqualifies Alaska from being a true candidate for Paradise). There's a pronounced interest in firearms and camo wear, and a singular shortage of womenfolk: those seeking a husband used to be warned, 'The odds are good, but the goods are odd.' It's scarcely a comfort zone, and has never really been a country for old men.

The lure of a Promised Land in the North continues to suffuse Alaska's modern history, and found its early expression in the writings of Scottish-born explorer John Muir, who mounted four expeditions there (the first in 1879) and was one of the first chroniclers of the more fertile south-eastern panhandle, along with the snowfields and icebergs of the colder coastline. Physically hardy, maybe a touch foolhardy, 'John of the Mountains' must have been a tiring (though seldom tiresome) companion. Imbued with an unsnuffable *joie de vivre* and breathy passion for the wonders of Nature as evidence of divine handiwork, he was also rare among early travellers in admiring the 'natural dignity' of several native tribes (notably the Stickeen and Tlingit) whose traditional way of life he described in prelapsarian terms: 'the very Paradise of the poets, the abode of the blessed', he wrote on his deathbed (1914), aware that even then the pristine places he remembered were being transformed detrimentally by the influx of other forces. Influenced by the Transcendentalists, he made this key observation: 'When we try to pick out anything by itself, we find it hitched to everything else in the Universe.' Today he is hailed as a pioneer conservationist: he had a glacier, a rose, a wren and a millipede named after him.

The subsequent impression of Alaska is strabismically skewed by two writers – Jack London and Robert Service – who furnished us with those abiding images of dog-sledding and the Gold Rush. Beginning in 1881, and peaking with the stampede of 1898, gold fever is associated with the Klondike (which was actually in Canada) but it affected the outsider's view of this entire region and its hectic culture, as tens of thousands trudged up the mountain trails into the interior in the hope of staking a claim, panning water and 'finding colours'. To most prospectors (nicknamed 'sourdoughs' for the live yeasty substance kept in a pouch next to the skin and vital for waffle mix), it turned out to be a fool's paradise of squalor and corruption – even Wyatt Earp lasted just one week as a marshal up in Wrangell – and many women ended up plying a different trade in the shacks of Dawson City's 'Paradise Alley.'

The prolific and widely travelled novelist Jack London had first-hand experience of the Gold Rush, and won an international following with *The Call of the Wild* (1903), his thrilling canid saga about the savage and the tamed, set in a brutal Northland's 'Arctic Darkness', where a treasured pet has to learn how to become leader of the pack. The author liked to refer to this book, which made him a fortune, as just 'a yarn', and baulked at critics who discerned in it themes of 'existential primitivism', though he enhanced this aspect of the search for redemption in the world at sixty below when his longer, complementary novel *White Fang* appeared in 1906. In its unsentimental portrayal of the rigours of wilderness and group dynamics, *White Fang* proved hugely popular, and influenced both Hemingway and Orwell. In Nancy Mitford's evergreen satire *The Pursuit of Love* (1945), her choleric Uncle

Matthew is asked by one dinner companion if he has read a certain fashionable new novel, and replies, 'My dear Lady Kroesig, I have only read one book in my life, and that was *White Fang*. It's so frightfully good I've never bothered to read another.'

Scottish-born Robert W. Service, known as the Bard of the Yukon, or the Canadian Kipling, remarkably didn't arrive in the Klondike until 1904, when its heyday was already over, yet his many ballads about cabin life in 'the Great White Silence' became best-sellers. Before he had even reached Dawson, he had invented a cast of characters, including Athabaska Dick and Claw-fingered Kitty, who peopled his demi-monde of gamblers, brawlers, bad hats, black sheep, loggers, drifters, remittance men and barroom roisterers. His volume *Songs of a Sourdough* (1907) alone earned him one hundred thousand dollars, and he later retired comfortably to a life on the French Riviera – yet his legacy of rough but Romantic life on the Last Frontier has never quite faded from the palimpsest on which subsequent Alaskan history has been overwritten.

'Hi yu muck-a-muck,' John Muir recalls the Tlingit shouting as they speared and hand-tailed salmon, while 'the water about the canoe ... was churned by thousands of fins into silver fire'. For the sophisticated, warlike Tlingit and the riverine Athabascans of the Interior, it was a matter of survival to harvest during a few summer months enough salmon from the various species to see the community through the winter, and it is hardly surprising these fish became objects of cultural significance, sometimes typified as successive flashes of underwater lightning. Certain tribes believed their ancestors had learned their catching methods and the attendant rituals from the trickster deity Coyote, and they deployed an ingenious combination of weirs and spears, gaffs and dip nets along with sweep nets several hundred feet long, woven from dune grass or nettle (from which we derive the European word 'net'). They regulated against overfishing – something the white men blithely ignored – and even had their own stream masters (*Héen saat'í*) to co-ordinate operations. Fish were smoked or wind-dried (they had not discovered salting), ground into a type of flour or preserved in bundles as pemmican. The salmon heads (but never the hearts) were fed to the dogs, without which the winter months were unendurable. The name of the 1960s band Three Dog Night refers to a native gauge of how many huskies you needed to keep warm as you slept.

The season's first fish was treated with great ceremony, its skeleton reassembled and restored to the water facing upstream to encourage its fellows to proceed. There would then ensue a celebratory 'potlatch' with feasting and dance. The animistic belief that animals possessed souls informed many native myths, and it was thought that salmon only gave themselves voluntarily to tribes who treated their annual self-sacrifice with respect (a notion some modern anglers might appreciate). There were many versions of the Salmon Boy legend – or Shin-quo-klah – a spoiled child who threw away a mouldy fish meal and was spirited away below the sea by the salmon people to their Great House to learn about their ways, joining their migration (some say in red canoes) and eventually being reunited with his family, to whom he imparted his insights. Becoming a great shaman, in old age he encountered a curiously transparent salmon with a red feather in its head, and the very moment he speared it he died, as this was his own spirit passing. Fishing was a crucial, vital business, refined and ritualised over many centuries.

But a few ancestral rites (or rights) were never going to stop the settlers from the south, once canning was invented, and runs of king and sockeye were harvested with ruthless new ingenuity – including the introduction, from the Mississippi region, of the scow-mounted 'fish wheel', which operated automatically in the current. There's a good account of this in James Michener's multi-epochal novel *Alaska* (1988), which charts the development of canneries in the 1870s, the immigrant workforce and the coming of mechanisation – especially the invention of the 'Iron Chink' in 1903, which largely replaced the Chinese workers who had previously gutted each salmon by hand. The industry peaked in 1918 with an annual export of tinned Alaskan salmon worth fifty million dollars, the motto of one factory being, 'We eat what we can, and we can what we can't.' With its familiar labelled red tins (originally lacquer-finished with linseed oil and red lead), canned salmon became the British and colonial housewife's versatile storeroom staple, a democratic and affordable foodstuff – though never (unlike its wild Atlantic counterpart) a luxury one. This level of slaughter was unsustainable, of course, even with the ingenious rebranding of the multitudinous humpy as 'pink salmon' in the 1930s, but the industry continues despite a setback in 1982, when fifty million tins were recalled after a botulism outbreak in Belgium. Today, strict catch quotas are in place for netting and there is mitigation stocking on some rivers, which in itself threatens the survival of natural wild stocks, and it is clear certain runs only exist on life support. In some ways it is a miracle that the underwater lightnings still flash as widely as they do.

I was determined to revisit the Alagnak, and in July 1999 I took the Doctor and my brother the Judge to join another of Arthur's parties. On our final morning, up at the Braids, we three decided for once to count the number of sockeyes we landed. Reaching a grand total of ninety-nine around noon, we let my father-in-law do the honours racking up 'the ton', while the Barnaby Rudge – still, despite a week's experience, scarcely able to believe the evidence of his own eyes at these sheer numbers – prepared some gin cocktails, and guide Ryan fired up his broiler for a bankside burger 'potlatch' that completed our session. (It was also to be the last time I ever fished with King Arthur, who died the following winter.)

The question arose: does this sort of fishing spoil you? Certain folk will assure you 'Pacific salmon don't really count' – supposedly because they don't quite look pukka and are too easy to catch, though that chap holding his abacus is likely also to chide you for putting mint in your Pimm's as opposed to borage. Of course, it's all relative. If your usual salmon-fishing experience is like a visit to some local small-town dance hall with only the vaguest hope of an occasional lucky 'pull', then an Alaskan trip more resembles a weekend in Paris with one of the late Madame Claude's finest *filles de joie*.

Ted Hughes began fishing in the Rochdale canal next to the polluted River Calder, and in his early years roamed the Yorkshire moors with his brother Gerald, imagining themselves to be Native Americans out on their hunting grounds. They both enjoyed reading Haig-Brown, and even fantasised about one day emigrating to British Columbia. More than once he described this childhood setting as a kind of paradise. Eventually, Ted did get to fish for steelhead in British Columbia – then in July 1980, with his son Nicholas, visited Alaska, chasing kings on the Gulkana (a tributary of the Copper river). He composed one of his most magnificent poems about it, 'That Morning'.

In his 1983 collection 'River', there are many poems that show what a complex matter fishing had become for him – images of sacrifice and healing, chivalry and sacrament, are caught up in syllabic swirls and strengthened by hard mineral vigour, particularly regarding the mystique of salmon and their 'epic poise'. Hughes establishes an interdependence between hunter and hunted: might it be that the fish (or even an entire waterscape) is actually the predator, capturing the angler as prey? There's that nightfisher in 'After Moonless Midnight' wading for sea trout and falling under the glamour-spell of the stream ('"We've got him," it whispered, "we've got him'"), and the preternatural experience of a take in 'Milesian Encounter on the Sligachan', when the Isle of Skye's water sprite 'grabbed the tip of my heart-nerve' – a moment of truth indeed. In a posthumously published interview, the poet discussed his experience of fishing in the Pacific Northwest and the sensation of something rising up (as it did in his early 'Pike') 'to grab you and *be* grabbed'. The angler is putting himself on the line. As few other writers do, Hughes transforms the way we feel as we address our quarry. I sometimes think he identified with fish because he himself was as stubborn, glamorous and mercurial as a salmon.

'That Morning' is imbued with a dreamlike quality that suggests a vision of the Day of Creation. He and his son (a fisheries biologist based in Fairbanks, sometimes known as 'King Nick' for his interest in chinook) find themselves wading 'Waist-deep in wild salmon swaying massed/ As from the hand of God', when two golden bears arrive and dive 'like children' to begin feeding on the kings. The scene is illuminated in a beautiful metamorphic sequence melding together the golden bears, gleaming chinook and the river itself (showing colours, perhaps, as it did for earlier prospectors of another type), along with the two men witnessing this transfiguration – the very act of observation hallmarking the moment: 'So we found the end of our journey./ So we stood, alive in the river of light/ Among creatures of light, creatures of light.' It is the culmination of a long quest, a homecoming, the completion of some circuit. Fishing very often generates this strange sensation of interconnectedness with the natural world, and in that last interview the poet (in essence endorsing Muir's observation) spoke of how it links you to 'The stuff of Earth ... the whole of life'.

Ted's memorial service in Westminster Abbey was held on 13 May 1999, and I arrived hotfoot from the Itchen as the mayfly hatches were reaching their apogee. The interior of the building resounded with the Laureate's own recorded voice reading the song from *Cymbeline* ('Fear no more the lightning flash,/ Nor the all-dreaded thunder stone') and I thought of his own luminous description of the Irish mayfly in 'Saint's Island': 'They hurl themselves into God.' Surely, he had followed them. In 2011, when his plaque was unveiled in the south transept of Poets' Corner, Seamus Heaney read Ted's verse, 'Some Pike for Nicholas'. The last three lines of 'That Morning' are inscribed on the green slate.

At a Bonhams sale of Hughes's family possessions – including Sylvia Plath's fishing rod – in 2018, I acquired a simple, full-height wooden lectern at which Ted occasionally stood to write (there's a little fish doodle scratched onto its upper plane). In my Scottish library it faces out across the trout loch, and sometimes I stand at it for luck, in the innocent hope that some scintilla of his poetic starshine, some vicarious inspiration immanent in the fabric of the joinery itself might somehow leach out and enlighten me. The lectern was made for him by Nicholas who in 2009, after years of depression, took his own life. In one obituary an Alaskan colleague noted, 'His writing was as clear as a grayling stream.'

*

Conceivably because I had bored them rigid on the subject throughout their adolescence, my own sons James and Tom both reached their thirties without ever having properly fished for salmon, so in 2017 I decided to remedy the situation by organising a traditional Dad and Lads trip to the Alagnak. Although I was toting my usual supply of lucky charms, we did not get off to a very auspicious start: practically all our gear was left by the handlers in Keflavik, and, four flights later, on our eventual arrival at the Alaska Trophy Adventures Lodge, I was informed there was no liquor bar in the camp (fortunately, our supply of writers' and artists' materials was safely in the hand baggage). At King Salmon there had been a delay as we waited in the hangar for another guest to arrive, a Mr Wayne Salmon – *Ex Alaska semper aliquid novi*.

Head guide Patrick raced us downriver the first morning in a jet boat, past the now-mothballed camp where the late King Arthur had held court, towards a long chum bar. Both boys (plus our sprightly American friend Todd) had fish on before I had even fumbled the fly out of my keeper ring. Things were finally coming together. For an entire hour we got in among the salmon: these were good, solid specimens up into double figures, some chromers and others already showing maroon markings on sooty olive flanks, as if ritually scourged, with a suitably hangdog demeanour – they have been renamed 'tiger salmon' in some places, but there's nothing feline about their behaviour once hooked. My sons adopted determined fighting stances, as if roping steers. It was not exactly subtle sport, but a magnificent curtain-raiser. James caught the first coho of the season, a good month earlier than usual (it was released, so there was no need for any skeletal rite). Pat soon had the boys performing snappy Circle-C casts, and before the sun came out and slowed down the 'bite', making the chums 'tight-lipped', there was already established what was to become a week-long claim that the insolent youths were easily out-fishing their paterfamilias. I could not have been happier. From our chalet's veranda that evening, we saw in the light of the setting sun a sow bear teaching her two golden cubs how to fish.

The lodge is located in the Upper Braids region, and fortunately a second run of sockeye was coming through the shallows so we had some almost indecently productive sessions in the side stream christened Costco, where we counted fish swimming past at the rate of forty per minute. 'Nice, Tom – you're just a fish vacuum!' veteran guide Justin would shout; 'You're red-hot today, James,' as once again the single-handers pulsed and their twangling lines unzipped the current. One morning at Black Hole we were to kill a limit for the camp freezer and in little more than two hours we had a leash of thirty fish on the stringer. A bear began ambling round the barbecue pit. We had palm blisters and line burns and finger guards, which you don't often see on Speyside. Justin called it a 'sockeyedelic' experience, and after that we genuinely lost count of the numbers: 'the fishing became almost incidental,' I wrote in my notebook (all the same, they had to drag me away from the Black Hole of Alaska). *Hi yu muck-a-muck* ...

'Ready to go out there and kick some more fin?' I enquired at breakfast of Buzz, the camouflaged and moustachioed Texan; from beneath his ball cap he grinned, 'Oh, baby!'

On our last day we took a long raft trip aboard two inflatables, side casting for rainbows and grayling with little egg patterns under sight indicators. It was a wonderfully quiet way of seeing the upper river as we drifted without engine noise, like being in a glider, the soft sift of the current hissing beneath our airy pontoons. At the end of a hectic week, the peace and sense of belonging illuminated the time spent on the water with my sons, despite the silent apprehension that nothing quite like this might ever happen for us again. We got out to wade some riffles with dry stonefly patterns for trout; Todd caught a small king from the Char Hole, and I got a humpy, so we racked up all five of the salmon species. The comeliest fish I captured was a Dolly Varden – a char named after a character with a fondness for fancy clothing in Dickens's *Barnaby Rudge*. Enrobed in quicksilver, its flanks were prinked with orange and pink stipple, and the anterior edge of the lower fins showed white details. She measured a mere half-inch short of the 2-foot trophy mark: I guess, like Arthur, I'll just have to come back and try again. I hope it's only a matter of time.

We had found the end of our journey. In Anchorage we bought ourselves some trademark brown rubber Xtratuf halibut fishermen's boots – the Tod's loafers of that city – and, by now looking a touch chum-like and no longer entirely fresh off the tide, we sloped along the sidewalks in search of strong waters. We lit upon Humpy's Great Alaskan Alehouse, a seafood joint where I was presented with an 'I Got Crabs at Humpy's' bumper sticker, which is certainly unique up in Perthshire. From there we made our way, swaying beneath the hanging baskets of marigold and lobelia, to a dive bar named Darwin's Theory, and thence to a lower joint where there was an open mike rock 'n' roll night. After one too many Jacks on the rocks (they were magnum Jacks), I had to be dissuaded from volunteering a rendition of Three Dog Night's 'Heaven Is in Your Mind'. I should know by now to stick to fishing.

Ah well, I reckon it's back to those local dance halls. Oh, baby.

19

OF PARADISE

*'Fishing is a world created apart from all others, and
inside it are special worlds of their own . . .'*

Norman Maclean, *A River Runs Through It
and Other Stories* (1976)

People entertain extravagantly various concepts of perfection, and the prospect of sublime or heavenly delights. To
the Victorian sybarite Henry Luttrell it was 'eating *foie gras*
to the sound of trumpets'; the spelunker dreams of his ideal
cave; the surfer chases that elusive wave. To Terry and Julie, a
Waterloo sunset over a dirty old river transported them safely
to paradise. The pursuit of a state of ultimate bliss is harmless
enough, so long as one remembers the inherent impossibility
of its attainment.

Although now degraded into promotional shorthand for travel and property brochurism, the notion of paradise remains a serious matter. Probably deriving from the Persian *pairidaéza*, signifying an enclosed garden or fenced orchard of the fertile kind so many generations of horticulturalists have yearned to recreate, a paradise may be earthly (such as our Garden of Eden) or a feature of the afterlife (such as the Muslims' Seventh Heaven, the Elysian Fields, or Tír na nÓg – the Celtic land of eternal youth, with its golden apples and enchanted well containing salmon). Much imaginative literature, from *Robinson Crusoe* to space operas, has been inspired by the fundamental fantasy of humans somehow being able to repair their fallen state and return to a life that is less tainted with fraudulence, savagery, subterfuge and conflict, represented by a distant Golden Age – pastoral, free of weapons and surrounded with a corona of peaceableness (that freedom from worldly cares that epitomises the moonwashed Himalayan courtyards of Shangri-La, for instance). That such an idyll is also implausible underlies most Utopian visions, not least Thomas More's original *Utopia* (1516), which describes an island where all seems communal and harmonious, yet its very name translates as 'No Place' (there is a misconception that it means 'good place'). But such a longing seems to be an enduring, essential human impulse. In Orwell's *Animal Farm* (1945) the raven Moses describes to the pigs a place in the clouds called Sugarcandy Mountain where linseed cake grows on hedges. In his play *The Birds* (fourth century BC), Aristophanes was already warning us about Cloud Cuckoo Land.

Anglers, being in some ways childlike, are often idealists. This can make us restless in our search for dream destinations, the prospect of heaven on earth – remote chum bars, forgotten carp pools within walled manor gardens, trout tarns hidden in the hilltop creases, secret lagoons where tarpon glide at dawn. Many of us pursue a lifelong quest for some silver fleece. I believe fishing is one way, however subconsciously, of seeking to re-establish our connection with the natural world, to gain readmission to a prelapsarian state from which – our faint race memory tells us – we were once banished. (Listen, sport, if you don't want to feel at one with the cosmos, could you move along to the next kiosk?) Other folk may pursue this source of restoration according to different lights – gardening, mountaineering and sailing can, I'm sure, furnish comparable feelings of replenishment – but fishermen try to retrieve their lost paradise with rod and line, knowing that despite our best efforts there will always be unfinished business (indeed, that's part of its point). You may never discover the zip code for Main Street, Paradise, but you can be quite happy pottering around its suburbs.

Most of the time, angling is a source of happiness. In times of stress – on a day when you feel the rats are pelting out of the cane field towards you, and the world sucks rocks – even the mere thought of fishing can sustain you with flashbacks to happier days, when for the time being your back was safely turned on all things troublesome ('Heaven Is in Your Mind,' as Three Dog Night remind us). Fishing may be a metaphor for life, I don't know, but certainly it can feel that way when your every overture is being spurned, or you are fast into an invisible tree during a nocturnal sea trout session, or the entire day is clouded with disappointment and loss. Much more frequently, however, it acts like a spiritual tonic. I have now almost six decades' worth of paradisiac memories to draw down whenever the immediate universe appears to be a disorganised, unprincipled mess, and you feel like a stranger in a strange land, in need of something clear and reliable for comfort. *Ubi piscis, ibi Arcadia*. Far away and long ago … crouching amid the burnside heather with my father, in the aftermath of his own fall from grace; tightening into an immense salmon as I kneel in a Canadian canoe; an island loch glazed with rainlight, and the crackle of water beneath the hull as we row back up the drift; deadbaiting for sharks at night, my line slipping away from a Bahamian dock where I'm lounging alongside a bottle of bourbon beneath a sky salted with stars.

The Christian tradition of earthly paradise depends on a nifty Creation fable by several hands which runs over three chapters of Genesis, and exudes an enduring mythic beauty. The question one should ask of myth (as with fairy tale) is not, 'Did this happen?' but 'Does this happen?' Yet I have always preferred to read the tale of Adam and Eve as if literally true, not along the lines of learned theologians such as Bishop Butler (who calculated that Adam named the animals on 4 October 4004 BC), but in terms of pictorial imagination. Some early cartographers, for instance, took to marking the location of the lost Garden on maps of Mesopotamia, and one specific Genesis detail that appeals to me is indeed potamic: 'And a river went out of Eden to water the garden; and thence it was parted, and became four heads' (Genesis, II, v10). These are generally accepted as being Pison, Gihon, Hiddekel and the Euphrates. Rivers feature in many Creation accounts (the floodplains of Dilhun, the four rivers flowing from China's Kunlun mountains) and I feel certain Eden's streams, which ran like nectar over pearl, were the site of mankind's first fishological moment.

Mediaeval clerics claimed Seth, Eve's third son, discovered the secrets of angling and mathematics and inscribed them on certain 'pillars of knowledge' that survived the Flood, but surely Adam must have been the Original Angler? The Bible doesn't say where he was when the serpent beguiled Eve, but I envisage him off fishing those virgin waters – the Fall of Man coincided with the first fall of mayfly spinners, perhaps – and when he came whistling back to their honeysuckled bower of bliss, there was his wife, hands on hips, suspiciously counting his ribs and having tasted of the Forbidden Fruit. (Apples were not native to the region, but may have crept into the account due to that similarity between Latin *malus*, evil, and *malum*, apple. Or maybe it was the tomato, a berry once known as the Love Apple, or *pomodoro*, 'golden apple'.) Of course, no fishing is ever quite what it was: Adam is said to have lived to the age of 930, and I'll bet each day and during the drear watches of the night – he spent a further four thousand years in purgatory – he regretted the loss of access to that idyllic river.

Milton omits this from his deeply influential epic, dictated in his sleep by a muse called Urania and transcribed by a series of dutiful amuenses (the poet had been blind since 1652), but he knew about the paradisian significance of fish. *Paradise Lost* was published in 1667 ('None ever wished it longer than it is,' growled Dr Johnson), the rights being sold to a printer for ten pounds – one wonders who Milton's agent was. Probably because the poet had been a fervent supporter of the Great Rebellion, his arch-villain is curiously charismatic and plausible ('The prince of darkness is a gentleman,' as Edgar reminds us in *Lear*), but when he is depicted in Book IV escaping 'the Stygian pool' and alighting on the Tree of Life, Satan has shapeshifted into a cormorant, exuding 'fishy fume' in contrast to Eden's 'odorous sweets' (a blind person's olfactory marker), and in that form he perches 'devising death/ To them that lived'. Because of their proximity to the royal aviary, Milton's lodgings near Petty France would have made him familiar with a bird long a byword for piscivore rapacity, nemesis of fishdom (the very creature that was emblem of the Christian god), a figure of ill omen in folk wisdom with its notably serpentine neck and possessed of a guttural, churring voice that is darkly foreboding. James I and Charles I both kept Norfolk cormorants there, trained, Chinese fashion, to go fishing for sport as a type of underwater falconry. Although he could no longer see them, Milton knew the heraldic pose, like a grotesque finial, these tenebrous birds adopt when drying off their wings after swimming – a black cruciform profile that offers a parody of the symbol of Christian sacrifice.

Phalacrocorax carbo (then known as a sea crow) was literally the embodiment of evil, a bête noire from the sulphurous pit. It is one of the Creator's truly diabolical inventions, along with spinach, haemorrhoids and trigonometry.

With a little spool of golden fly-tying thread beside it, a bust of William Blake (modelled on his death mask) looks down from the mantelpiece in my library. The visionary artist – who, at the age of four saw the face of God at his bedroom window and later beheld a human figure covered in scales at the top of a staircase – identified strongly with Milton's sympathy for the devil, and spent so much time obsessively illustrating the poem for his 1810 edition that his long-suffering wife Catherine remarked, 'I have very little of Mr. Blake's company. He is always in Paradise.' His own writing frequently engages with versions of Eden and the unrelenting spiritual search for reattaining a state of grace, and in his late, lengthy and obscurely prophetic work *Jerusalem* (not the popular hymn, the lyrics to which appear in his *Milton*), he has this verse: 'I give you the end of a golden string,/ Only wind it into a ball:/ It will lead you in at Heaven's gate/ Built in Jerusalem's wall.' The clue to heaven here is a celestial handline, a lightning thread that leads you up to Paradise.

When we consider the heavenly afterlife – the one to which Christ on the cross referred, 'Verily I say unto thee, Today shalt thou be with me in paradise' (Luke 23) – there is a common motif in many cultures of a place of reunion, of the joining up and binding together of the broken threads of your earthly existence. A notable depiction of the meeting of loved ones appears in the elegant altarpiece panel by Giovani di Paolo di Grazia, *Paradise* (1445). Its foremost literary expression is found in Dante's *The Divine Comedy*, written during the previous century and published in 1472, which is another example of a visionary classic that has influenced many who have not even read it.

Beginning on Good Friday 1300, the poet famously loses his way in a dark wood, *selva oscura*, and finds himself on a conducted tour of the afterlife with Virgil as his spirit guide. The 'Inferno' has captured the imagination of subsequent readers because of its phantasmagoric details, as is the way of things, and its lurid descriptions have defined many aspects of what Hamlet called The Other Place. The third canticle, 'Paradiso', which culminates in a vision of beauty transfigured with creatures of light among the stars, is perhaps a little too esoteric for the casual reader, though it does contain a nice reference to fishing for the truth, and how that is an art (*'chi pesca per lo vero e non ha l'arte'*) (canto 13, l. 121), and a memorable image of converging angelic souls rising to greet the pilgrim like fish gathering towards food in a stew pond (*'peschiere'*) in which the water is still and pure (*'ch'è tranquilla e pura'*) (canto 5, l. 100) – significant attributes of a heavenly element. But for a description of Eden we must look to the middle canticle, 'Purgatorio', which describes an allegorical journey through limbo, the celestial halfway house, God's keep-net.

Set in the southern ocean, there are seven levels to Mount

Purgatory and at its summit is the Garden of Eden, characterised by sweet breezes, bird call and perpetual spring. Adam and Eve are serving out their time there while atoning for the Original Sin, and the narrator's pilgrimage brings him to the bank of a river which cleanses all memory of transgression. Here he is reunited with the spirit of his great, unrequited love, Beatrice, who baptises him in the sacred water – another place of safety and joy with a river running through it.

Virgil was a significant choice of guide, as he was the celebrated poet of a pastoral tradition that explored the image of Arcadia (the memento mori phrase *Et in Arcadia ego* is attributed to him). In Greek mythology, Arcadia was a remote, bucolic region immaculately rural, wild and unspoiled, and it became a psychogeographical template for many explorers in pursuit of lands of plenty. The discovery of the 'New World' from the time of Cabot onwards attracted many descriptions of the wonderful abundance of this promised land, in particular its practically unfished waters. European boats were soon plying the Grand Banks for dense shoals of cod, and early settlers around Chesapeake Bay reported such biblical hauls of shad, lobster and sturgeon that they ran out of salt supplies for preserving their catch. Surplus fish were used as fertiliser, and sturgeon were dried and stacked like cordwood for fuel. It was not long before this divine bounty was spoiled by the law of diminishing returns but the New Found Land and a vast dream of colonisation inspired the Renaissance imagination, and its reforged sense of paradises yet to be revealed – though, as Shakespeare, among others, demonstrated (for instance with Prospero's embattled island), our perceptions of paradise are likely to prove delusory.

In the same era that Milton was hymning the loss of human innocence, and Walton was confecting the Arcadian delights of rural Albion, there was a vogue for poems celebrating the virtues of the country house retreat, often in Edenic tones and with paradisal allusions. One of the finest is 'Upon Appleton House' (1651), Andrew Marvell's paean to the Fairfax family's Yorkshire estate through which flowed the Wharfe, where the poet happily angles with some success ('While at my Lines the Fishes twang!'). The river is specifically serpentine, and winds through an orchard complete with 'luckless Apple'; since this tidal stretch of the Wharfe occasionally floods the fields, Marvell records how salmon and pike become stranded in strange places when the inundation recedes. This long poem culminates in an elaborate metaphysical conceit, wherein at the end of the day the salmon fishers hoist their leather canoes above their heads and this is likened to the coming of night: 'Let's in: for the dark Hemisphere/ Does now like one of them appear.' Rivers and fishing were commonly accepted as key paradisal motifs.

A river runs through Kubla Khan's Xanadu, also, in the narcotically fuelled fragment by Coleridge, who, when he fell asleep, had been reading an account of Marco Polo's possibly apocryphal visit to the bloodthirsty Tatar's retreat. Although the warlord has attempted to create a paradise (the milk of which a visitor may drink), it is a flawed one: the cavernous Alph, which doesn't sound very attractive to an angler, disappears into a 'sunless sea', and the peak where the pleasure dome is constructed is (in the manuscript original) Mount Amara, where Milton located his false paradise. The Romantics were much taken with such concepts of the sublime and idyllic: in Coleridge's diffuse *Notebooks*, he describes a beck on the fells, 'which looks like a paradise' (1802), and later muses, 'I would glide down the rivulet of quiet life, a Trout!'

Several times, Proust described as a paradise the fictionalised provincial town of Combray, where his autobiographical narrator spends his childhood, and fish feature surprisingly often in his recollection of lost time: float-fishing on the water at Tansonville just before he meets the beguiling Gilberte Swann, boys trapping minnows in glass jars on the Vivonne river, and his early fantasy that the elegant Oriane, Duchesse de Guermantes, will take a shine to him and stand forever by his side, fishing for trout. This aristocratic cynosure survives the entire novelistic sequence, and in the final volume is arrayed like a salmon, with fins of black lace – an old 'sacred fish'. Perhaps he was right that the only idylls worth pursuing are *'les paradis qu'on a perdus'*.

The identification of rivers with Arcadian iconography continues in modern Western literary tradition. Richard Brautigan (who once lived in Paradise Valley, Montana) literalised the break-up of the American dream in his chapter, 'The Cleveland Wrecking Yard' (*Trout Fishing in America*), where sections of dismantled Colorado river are stacked in lengths for sale at six dollars fifty per foot (there were waterfalls stored upstairs in the used plumbing department). In another deadpan pastoral satire, *In Watermelon Sugar* (1968), his Adam and Eve protagonists burn 'watermelon trout oil' in their lamps, with the climax coming at a funeral dance in a fish hatchery. Cormac McCarthy's solemn novel *The Road* (2006), a post-apocalyptic portrait of a world irretrievably lost, reflects in its final paragraph on how the mountain streams once held brook trout finning softly in the flow of the current.

The loveliest exploration of this theme is surely Norman Maclean's strongly autobiographical novella *A River Runs Through It*, set in the Montana countryside of his 1930s youth, and an elegy for an earlier time. I take 'What a beautiful world it was once' to be the book's essential sentence, although its characteristically wry opening paragraph about there being no clear line in their Presbyterian family between religion and fly-fishing has become the more celebrated. There is so much to be admired in this Western story (it truly passes the rereading test) that it seems reductive to claim it is about the possibility of saving graces and redemption through the achievement of beauty, though both of these themes surface occasionally like the bulge of water over a boulder in one of the Big Blackfoot's challenging canyons, where the narrator's younger brother Paul – whom he loves, but cannot understand – transforms his fishing by a technique he calls 'shadow-casting' (to simulate the onset of an insect hatch) and even on his path to destruction arouses the admiration of his anxious family. It is partly an essay in forgiveness, and, as I recall from my own relationship with my father, there are lessons to be learned about how to forgive but also about how to be forgiven.

It was their minister father's stated belief that 'man by nature was a mess and had fallen from an original state of grace', and it is through the medium of the water that the Maclean men seek the currency of order and congruence – 'Something within fishermen tries to make fishing into a world perfect and apart.' This world – through which the river flows, over rocks that go back towards the earliest of times, where you can read in the water its ancient message of continuity – is a sanctuary threatened with defilement by the interloper Neal, a bait-fishing brother-in-law who brings the town whore Rawhide into their already fragile Arcadia, yet the brothers manage one final day on the water before Paul's demons bring about his destruction, and 'on this wonderful afternoon when all things came together' a precious reunion is achieved. You need never have even held a fishing rod to feel the hydraulic pulse of this beautiful myth.

Fish have featured in so many tales of fertility and regeneration that they have become as complex and ancient a motif as the serpent. Creation stories from around the world involve fishy origins: the Japanese Ainu believed the world was balanced on the back of a cosmic trout, its movements creating earthquakes and tsunamis, while a colourful Polynesian myth describes how Maui drew up the islands of Oceania on a fishing line made from his sister's pubic hair. Swallowing and regurgitation by fish or cetacean is another common image in maritime cultures, echoing the way the sea marvellously throws up the swallowed sun each dawn. Theriomorphic deities with scaly attributes include Vishnu (one of whose incarnations when he came to save the world was the half-fish Matsya), and the Aegean Dogon – not to be confused with Dagon, the Sumerian corn god. Japan's cheery Ebisu, one of the seven gods of good fortune, is habitually cradling a carp. In areas where astrology was potent, the double fish zodiac sign now known as Pisces symbolised both the end of the old year and the beginning of the next, which the Magi would have recognised as they journeyed towards the winter Nativity. By the time Christian iconography was being shaped, the fish was already an intricate mosaic of emblems.

The Bible abounds in watery stories, of course, from the Flood (Ham and Shem must surely have trolled behind the gopherwood Ark, or else Noah's family were reduced to being vegetarians), through Jonah, Leviathan and the New Testament miracles – the draught of fishes, the feeding of the five thousand (those personal food wallets which were shared around probably included locally caught tilapia) and the taking of the tribute money from the mouth of a fish at Capernaum, 'Go thou to the sea, and cast an hook' (Matthew 17, v27). As well as the John Dory, several species are still said to bear the imprint of Peter's thumb as a consequence; others are popularly held to show biblical signatures, including an American catfish variety (*Arius*), the otoliths of which resemble the dice cast by soldiers at Golgotha, and I once met a woman in New Brunswick who gave me the desiccated interior of a lobster skull which she swore was a sculptural likeness of the Blessed Virgin Mary (after a fair bit of Canadian Club, I could just about make it out).

The Apostles were designated fishers of men, and mediaeval traditions held that fish were even an essential part of the Last Supper – sometimes Judas is depicted stealing a fish and concealing it beneath the Eucharistic napery. Also adopted as the secret symbol of early Christian disciples (who in certain circles were referred to as *Pisciculi Christianorum*, 'little fishes'), a fish sign in times of persecution could easily be drawn in the sand or dust by the faithful with two arcs of their foot, a simple outline of the *'vesica piscis'*, an ideogram representing submersion in the waters of baptism and, in its resemblance to the Greek initial letter alpha – itself a vaginal icon – the connotations of rebirth and a new beginning. The Greek *ichthus* also functions as an acrostic for the names of the saviour – Iesus Christos Theou Ysos Soter – and the sign has enjoyed a recent afterlife in various bumper sticker decals for followers of Darwin, Rasta or *Star Wars*.

The fish is therefore a polysemic symbol, as Carl Jung explained at length in his study of archetypes. His book *Aion* (1951) is subtitled 'Researches into the Phenomenology of the Self' and contains a chapter called 'The Fish in Alchemy', which I cannot pretend to have understood completely, but it traces the evolution of the fish as an image of wholeness back to the worship of the Babylonian god Oannes, whose priests wore fish skins (as, in Jewish tradition, did the faithful in Paradise).

He proposes a connection between the cult of the Virgin Mary and a half-fish maternal deity of the Syrians named Derceto-Atargatis, from which eventually emerged symbologically the concept that Christ was both bait and hook for God's rod to subdue Death in the form of the angled Leviathan. He notes along the way the ambivalence of attitude to fish in various cultures – widely caught and valued in some, but abominated as taboo in others, or else venerated: certain Egyptian priests would not touch the barbel-like medjed (*Oxyrhynchus*, or 'elephant fish'), for instance, as it was believed to have consumed the phallus of Osiris (god of the resurrection) when he was dismembered by his treacherous brother Typhon. This sort of analysis will never help you catch more fish, I admit, but at least we now have an idea of what bait to use for barbel.

Partly due to its tenuous association with the apostolic métier, angling as a recreation gradually acquired an image of probity and wholesomeness from the Middle Ages until, perhaps, its mass popularity in the Victorian period (the young Norman Maclean assumed from his father's approval that on the Sea of Galilee they were all fly-fishermen). Unlike the sport of kings, it was quietly pursued by monks and men of the cloth, and never attracted those dissolute figures of the hunting world. Many divines were enthusiastic anglers as well as being fishers for men's souls: the poet Donne, the Revd Robert Nobbes (who wrote the first book on trolling, in 1682), Dean Swift, Bishop Browne, Charles Kingsley and Canon Greenwell – after whom the Glory trout pattern was named, and who was still fishing at ninety-two. It may not appear such a sanctified pursuit today, but there was a time when angling was perceived as positively saintly.

Once, people were intensely familiar with the images of individual saints, whose intercession from their elevated place in Paradise was necessary for one's own salvation in the afterlife. Even if you could not read, they needed to be identifiable in church carvings or illustrations, from a visual vocabulary of motifs and attributes. Thus, Gertrude had a tapir, Benedict a broken sieve and Juan de Dios a pomegranate. In more grisly fashion, many held the instrument of their own martyrdom – Agatha displays the shears with which her breasts were cut off, Lucy a plate bearing her gouged eyes, Elmo the windlass onto which the torturer wound his entrails, and Lawrence the gridiron on which he was roasted, asking to be turned over when one side of him was cooked – but, enough already.

There are at least thirty-four saints in the canon with fish as emblems, ranging from Corentin of Quimper (who caught a fish in his bucket) to the golden eel of Spiridon. Anthony of Padua, a Portuguese nobleman, preached a sermon to the fishes at the mouth of the River Marecchia, and they bowed their heads in devotion. Amelberga of Temse was molested by Charlemagne (who broke her arm) and she escaped on the back of a convenient sturgeon. The Swabian monk Benno of Meissen retrieved the keys to Elbe cathedral from the mouth of a fish, and – a common motif in folk tales – a ring was rescued in similar fashion by Glasgow's Kentigern (who died in his bath aged 185). Wilfred, Archbishop of York, taught the

Saxons to angle during a famine, and the Roman missionary Mellitus was shown by a Thames salmon-netsman named Aldrich where, in a vision, he had seen angels building a church and there he accordingly founded Westminster Abbey. The patron saint of fish is Neot, a diminutive Cornish monk and tutor to King Alfred (sometimes said to have stood just 15 inches tall, perching on a stool to celebrate Mass); he was sustained by two miraculously self-regenerating salmon, and left as a relic his comb fashioned from pikes' teeth.

For the angler, pride of place in all hagiography must go to Bishop Zeno of Verona, a fourth-century dark-skinned African orator and scholar famous for consecrating virgins to the veil, later patron saint of the fish hook. His legend has it that he was fishing in the Adige river when the Emperor Gallienus summoned him to exorcise his daughter and rewarded him with a golden crown, which the future saint broke up and distributed to the poor. He is depicted with angling accoutrements, sometimes with a fish dependent from his crozier. Some maintain he was martyred by Julian the Apostate (an early 'anti', perhaps). In 588, when the river burst its banks, the floodwaters miraculously refused to enter the Basilica where Zeno lies entombed in the crypt, his face encased in a silver mask. If I ever make it to Paradise, I know whom I'm going to ask for as my gillie.

You sometimes hear a particular species – milkfish, dorado, steelhead – being referred to as a piscatorial holy grail, but in fact the Grail legend itself involves a mysterious angler.

A popular subject for mediaeval narrative poems, the quest for the Grail followed the fortunes of various knights in search of the chalice from the Last Supper (though in some iterations it is actually a platter for serving fish). In an early script version of Monty Python's film skit on the topic, this ultimate sacred relic was to be found in a specialist Grail department in Harrods, but more traditionally it is conserved in the castle of the curious Fisher King.

The earliest surviving poem in which he features is *Perceval*, an unfinished romance in Northern French by Chrétien de Troyes, dating to the 1180s, with a sequel trilogy by Robert de Boron in the following century (Wagner's long opera *Parsifal* – a favourite of Hitler's – is based on these stories). Details of the Arthurian knight's encounters with *le roi pescheur* vary, but to synthesise them we find a curiously allegorical figure representing Man in his fallen state. He languishes in a barren land, where a javelin wound between his thighs ('*parmi les cuisses*') has robbed him of his masculinity so that he can no longer ride, and he has resorted to fishing in the river, a process whereby he may somehow heal himself and be restored (the fish here functioning as a female symbol of fecundity) and his kingdom will be reinvested with its former fertility. In some versions he is five hundred years old, and is identified variously as Brons, Parlan, Anfortas (the Unwell One) or St Joseph of Arimathea, who used the chalice to collect blood flowing from one side of the crucified Christ. Another precious talisman kept by the Fisher King is a bleeding lance or spear, specifically the one used by the blind centurion

Longinus at Calvary. Unable to produce an heir, the guardian can never be released from his role until a questing knight is shown these relics and asks him a particular question, whereupon he can reveal the secret of Christ's last words from the Cross: Perceval eventually does this, and the King instantly recovers – 'fit as a fish'.

As Jessie L. Weston noted in her seminal book *From Ritual to Romance* (1920), the lance and cup are 'sex symbols of immemorial antiquity and world-wide diffusion', as befit a synthetic fertility legend – but why is the stricken King an angler? I believe it is because he becomes a readily recognisable embodiment of dedication, persistence, patience and solitariness – qualities that have been associated with angling since early times – and he is also a trope for the human need to haul up something from the depths of the psyche, *de profundis*, into the revelatory light of religious belief.

The mysterious King has inspired many poets, from Wordsworth (who, walking with Coleridge in 1800, comes across a lone and lean fisher figure at Grasmere), to the listless, wandering Connemara angler of Yeats, but the classic application of the myth is in T. S. Eliot's *The Waste Land* (1922), which in its final movement has the Fisher King trying to reassemble his fragmented landscape and listening for the voice of the Thunder presaging rain. In fiction, there is a striking resemblance in Captain Ahab, who perceives himself (in one of his more Lear-like fits), 'as though I were Adam, staggering beneath the piled centuries since Paradise. God! God! God! – crack my heart!' He too has been mutilated, when once found unconscious in Nantucket after a fall, 'his ivory limb having been so violently displaced, that it had stake-wise smitten and all but pierced his groin'. He is reduced to a solitary existence with fading powers, in single-minded pursuit of a 'fish' in the hope of some kind of salvation. Bernard Malamud's innovative first novel, *The Natural* (1952), uniquely combines the world of baseball with the pursuit of a holy Cup, though Anthony Powell's *The Fisher King* (1986) makes disappointingly slack use of the legend.

Finally, Hemingway – who, in a 1925 letter to F. Scott Fitzgerald, wrote that his idea of Paradise would involve a trout stream that no one else was allowed to fish in – afflicted several of his leading characters with genital disfigurement similar to the Fisher King's. In 1919, when serving with the Red Cross on the Italian front, Papa himself had received a shrapnel wound between his legs and gives both Jake Barnes (narrator of *The Sun Also Rises*, 1926) and Nick Adams (another alter ego who features in a couple of dozen short stories) unspecified injuries, the emasculatory consequences of which they seek to exorcise by fishing. Barnes, an expatriate American in Paris, goes off to the Basque Country in pursuit of trout, while in the earlier back-country masterpiece 'Big Two-Hearted River' (1925), the maimed Nick, a young soldier back from the Front, seeks fulfilment and peace by fishing (as Hemingway himself did on returning to Michigan), though by concentrating on the minutiae of grasshopper fishing, the story misdirects attention from Nick's emotional conflict. The novelist did suffer temporarily from impotence in Paris, resorting to electro-therapy and glasses of calves'-liver blood, but claimed he was restored to full working order only when he converted to Catholicism.

The thirteenth-century Perceval interpretation *Parzival* and its prequel *Titurel* by Wolfram von Eschenbach happen also to contain the first European reference to fly-fishing. The chivalric knight Schionatulander, on his trail of the Grail and trying to impress his ladylove (who, alas, continues to read her book – a scene indisputably taken from real life), uses a *verangel*, or 'feathered bait', to catch perch – and, fittingly, grayling.

At the end of his long and surely blameless earthly existence as a divorce specialist, Jasper Silkstone QC duly presented himself at St Peter's check-in desk, and was directed down a charming lane lined with fragrant hedgerows, where dunnocks piped agreeably in the morning sunshine. He found himself already attired in a new fly-fishing vest and hip waders, though they had let him keep his favourite Turnbull & Asser cravat. With his habitual courtroom saunter, he arrived at a perfectly thatched hut with a flawless stretch of chalk stream purling before it. On the gleaming gravel patches between swaying tresses of viridian weed he could see several good brownies steadily gulping down a mid-morning hatch in rise forms the size of a Garrick Club dinner plate. Plumply, he surveyed the scene (was that a cuckoo? how delightful) – and, not for the first time, our lawyer congratulated himself that things were working out nicely, just as they should. A nattily attired young river keeper with a goatee beard respectfully presented him with a Leonard rod fashioned from exquisite, honey-coloured cane, matched to a bespoke titanium reel. 'Well,' exclaimed our lawyer, who could hardly wait, 'this certainly does look heavenly.' His companion gave him a watery smile. 'Alas, sir, no. You see, this is The Other Place – and although I may be Lord of the Flies, down here they never let us have any hooks.'

20

THE TROUT DON'T RISE IN
GREEN WOOD CEMETERY

'A man who does not exist,/ A man who is
but a dream.'

W. B. Yeats, 'The Fisherman', from
The Wild Swans at Coole (1919)

Through the open window of my bedroom I can feel the salt
breath of the sea coming in off the bay. I arrived on the island
last night in the dark, and I still don't quite know where we
are. The mansion sits back from an estuary, where the tide is
low and there is a series of sandbars with ambling shorebirds
and a rising breeze. Beyond, the open sea is as grey as a whet-
stone, the morning sun sharp upon it.

In front of the laird's house is a rough, mossy lawn and to its left a short, steep gorge of dark boulders descends to the bay, with a meagre stream running through it. There's a series of small holding pools, sunken now and waiting for rain – nice-looking spots for a worm tackle, though I'm not sure that's allowed. There has been a long summer's drought – weeks of sun searing the deer grass and heather, and no fresh sky water to let the fish run up the burns to the safety of the lochs. But today it seems there might be a change in the air, where a few clouds are gathering, baggy with moisture, out towards the west: perhaps at last the burning spell of the heatwave has been broken.

I am not sure who else is here at the moment, but I gather it's a mixed party of family and friends, with some others I have never met – a full house, though I haven't seen them yet. On the sideboard in the dining room is a mighty ham, and a hotplate with a dish of haddock kedgeree and a platter of fried eggs, their yolks a deep orange from the hens grazing along seaware. A glazed white bowl holds blaeberries off the hill, and some red gooseberries gathered from the small garden behind the house, walled off against the sheep. There's also a tureen of porridge, with a jug of golden cream – but I think I'll stick with my cornflakes. I finish breakfast alone: my stay here is just for three days, and I'm keen to get out and make the most of it.

Although I've never been here before, aspects of the place seem familiar. On the flagstones in the hall is a great bear-skin rug, a trophy from the Northwest Territories, and on the walls hang shields and crossed lances from the Sudan;

there's an antimacassared armchair, a pair of heavily upholstered Chesterfield sofas, an antler chandelier, and a tapestry
of Atalanta above the fireplace. On the round central table
someone has placed the Admiral, replenished with island
malt, and next to it sits a tweed trilby with a jay's feather
in the hatband. (Was that Mr White I saw out of the corner
of my eye, disappearing behind the green baize door, in
a slipstream of Vitalis?) Outside the tackle-room there's a
venerable Land Rover, with my favourite 10-footer clipped
to the roof, and inside sits Myrtle, the ashen-muzzled black
Labrador belonging to Alec the gillie. It's a long while now
since last I saw him: we used to fish together when I was
a teenager, before he crossed here from the mainland, but
he still looks remarkably youthful – burly, soft of speech,
blue-eyed. Today, I have him all to myself, and there is much
talk of days gone by, as the car makes its way up the dusty
track to the Loch of the Hazels, a long water at the end of
the whole system, where there's our best chance of catching
a proper wind.

At the boat house he says, 'Let's have a look in that box,
now,' and I open up the marbled oxblood Neroda my uncle
gave me, so he can choose a fly. Alec pinches out a double-
hooked pattern – a black hairwing with sparse purple hackle
and a body closely wound about with golden twist. 'Well,
that's maybe the boy for a fish. And a Pennell for your dropper, in case there's sea trout on the move.' He tilts back his cap
and bends to unshackle the wooden boat. There's no need for
the cattle feed scoop that serves as its bailer – the blue boards

on the bottom are bone dry. He manoeuvres us out between the waterlily beds and once we make it into the main arm of the loch there is more breeze than I expected. Alec looks seaward over his shoulder: 'Well, it's coming now, right enough.' The wet weather is bearing down towards us over the brown hills like a great grey bird, and things look promising as we begin our first drift. The dog drowses on the painter coiled in the bows. Decent waves begin to slap our strakes, and the boatman hunches over his oars, deliberate, patient, intent. Nothing moves to my flies, however, as for two hours we explore the long, indented shoreline. Then, when we reach the bay at the bottom of the loch where a thin burn comes down below some mountain ash, he sidles the boat closer into the bank. 'There,' says Alec, suddenly.

Quite close in against the rocks a dorsal fin and dark back arc out, I see the curl of a salmon's tail, and for a wonderful moment we are connected. It swerves out of the water, lugs away from us, and the hookhold fails. 'We've lost him,' says Alec. 'Too quick on the strike.' He turns now to row us back up against the rapidly swelling weather, and the first rain sweeps across the surface of the loch. Surprisingly soon, it has become a deluge. The sky settles its wings over us, and there's a livid, other-worldly light. The wind turns tumultuous and squally, fletching the wavetops with spindrift. 'Looks like we've had our chance,' says Alec, straining as our bows part the plunging water. Lightning briefly shivers and glims, then a trundle of thunder. 'We've lost him,' the loud air seems to echo, as we head for the shelter of the little island where a lunch hut looks welcoming amid the silver birches and the tall ling heather bending hugely beneath the gale.

Along the walls of the bothy are traced outlines of notable fish from earlier seasons; the shelf holds empty wine bottles plugged with candle stubs, and scattered corks where retired flies have been impaled by their broken hook points. Myrtle stretches in front of the stove, giving off an aroma of damp earth. There are good things in my picnic bag – a flask of hot consommé, a pasty, an apple, some bitter chocolate – but first I unpack the Admiral, and pour a drink for Alec to take from its silver cup. The wind bellows across the chimney top during our meal, and afterwards I feel in the pocket of my tweed coat and discover a packet of cigarettes (I seem to have taken up smoking again), as Alec lights his pipe and a bluish curl of thin smoke snakes into the air.

Conditions are now tempestuous, and we must start our long row back up to the head of the loch, to avoid being stranded down here. It's a good mile or so, and we take an oar each. At last, we make it through the narrows and into the lee of the top bay. Alec hauls the boat up the wooden rollers of the slipway so it's clear of the storm, and bids me try a couple of casts off the shore before we go. It's hard to get my line out towards the lily bed, and when the fish takes I never see him, so there's no time for me to make any of my usual mistakes – when he begins to churn the surface in his own storm of broken water it is clear he is large and dark, then he sounds in the direction of the tangled stems, shaking his head and lunging for safety, but never making a run for the open bay. Eventually we have him in the net. He's been up a good while, probably part of the sparse spring run, his flanks now tarnished by months in the peaty water, with none of that mauve sheen from earlier days. I remove the fly from his scissors – his teeth have almost destroyed the dressing, and the golden thread has unravelled. His body bears the marks where, months ago in the estuary, scales were raked away by an otter, or perhaps a cormorant. I wouldn't dream of ending his life now and, stooping in the shallows, I restore him to the water, cradling his head upwind until he revives and glides wearily out of sight. He has saved my day.

Wipers clattering against the frenzied weather, we head for home. The hills are streaked milky white where water is gathering down the crags, and when we make the final descent towards the house, with windows lit and teatime peat smoke slanting from the chimneys, it seems a real haven. We stop on the bridge to inspect the outflow from the Laird's Loch, which doesn't take long to fill in such a downpour. 'She's just about rising, anyway,' Alec says. The tide is out as we scan the bay. 'Only a matter of time, now.'

'Until tomorrow, then,' he says, as we reach the back door. 'Yes,' I reply, 'tomorrow's the day.' I turn to say goodbye to my companion, but already he has disappeared.

By dinner, the storm has passed through and before our evening *ceilidh* I can't resist the temptation to slip outside and look at the gorge. The stream is weltering down, bouncing off the rockfaces, turbid and gouted with foam the colour of buttermilk. I can smell the stain of peat it is carrying, and so can the fish that are now thronging the sea pool, responding to its freshness, eager for the last stage of their journey. In the glorious golden light, a fish hurls the blade of itself into the air and slices back through the darkening tide. I turn once more for the house, where faint island music – both restful and rousing – is coming from deep within. Across the bay and into the horizon a path of honeyed brightness stretches from the sinking sun. Tomorrow will surely be perfect.

Sources and Further Reading

Where possible I have given details for first published editions, but in many cases there are plenty of more affordable reprints (I examined one copy of a 1653 Walton that was on sale for £75,000). In some other instances, I have cited the copies from which I have actually worked. Unless otherwise specified, place of publication is London.

When it comes to subtitles, I have been selective in the interests of brevity. An extreme example to demonstrate the rationale behind this is Richard Franck's book, the full title page in my copy reading: *Northern Memoirs, Calculated for the Meridian of Scotland. Wherein most or all of the Cities, Citadels, Sea-ports, Castles, Forts, Fortresses, Rivers, and Rivulets, are compendiously described. Together with choice Collections of various Discoveries, Remarkable Observations, Theological Notions, Political Axioms, National Intrigues, Polemick Inferences, Contemplations, Speculations, and several curious and industrious Inspections, lineally drawn from Antiquaries, and other noted and intelligible Persons of Honour and Eminency. To which is added The Contemplative and Practical Angler, by way of Diversion. With a Narrative of that dextrous and mysterious Art experimented in England, and perfected in more remote and solitary Parts of Scotland. By Way of Dialogue. Writ in the year 1658, but not till now made publick, By Richard Franck, Philanthropus. Plures necat Gula quam Gladius.*

Andrews, John, *For All Those Left Behind*, Mainstream Publishing, 2002

Ashley-Cooper, John, *A Salmon Fisher's Odyssey*, H. F. & G. Witherby, 1982

Austin, A .B., *An Angler's Anthology*, Country Life Ltd, 1930

Badham, Revd C. David, *Ancient and Modern Fish Tattle*, John W. Parker and Son, 1854

'BB' (Denys Watkins-Pitchford), *The Fisherman's Bedside Book*, Eyre and Spottiswoode, 1945

Beazley, David, *Images of Angling: Three Centuries of British Angling Prints*, Haslemere: Creel Press, 2010

Bergman, Ray, *Trout*, Philadelphia: Penn Publishing Company, 1938

Berlin, Sven, *Jonah's Dream: A Meditation on Fishing*, Phoenix House, 1964

Berners, Dame Juliana (attrib.), *A Treatyse of Fysshynge with an Angle*, from *The Boke of St Albans*, Westminster: Wynkyn de Worde, 1496

Best, Elsdon, *Fishing Methods and Devices of the Maori*, Wellington, NZ: Te Papa Press, 2005

Boosey, Thomas, *Anecdotes of Fish and Fishing*, Glasgow: Hamilton, Adams & Co., 1887

Brandt, Andres, von, *Fish Catching Methods of the World*, West Byfleet: Fishing News Books Ltd, 1972 edn

Brautigan, Richard, *Trout Fishing in America*, San Francisco: Four Seasons Foundation, 1967

Brennand, George, *Halcyon: An Angler's Memories*, A. & C. Black, 1947

Briggs, Ernest E., *Angling and Art in Scotland*, Longmans, Green & Co., 1908

Browning, Mark, *Haunted by Waters*, Athens, OH: Ohio University Press, 1998

Buchan, John (ed.), *Musa Piscatrix*, John Lane, 1896

Buckland, Frank, *Log-Book of a Fisherman and Zoologist*, Chapman & Hall, 1873

Buller, Frederick, *Angling: The Solitary Vice*, Machynlleth: Coch-y-Bonddu Books, 2000

Caine, William, *Fish, Fishing and Fishermen*, Philip Allan & Co. Ltd, 1927

Callahan, Ken, and Paul Morgan, *Hampton's Angling Bibliography 1881–1949*, Ellesmere and Machynlleth: Three Beards Press, 2008

Chalmers, Patrick, *At the Tail of the Weir*, Philip Allan & Co. Ltd, 1932

——, *The Angler's England*, Seeley, Service & Co. Ltd, 1938

Chatham, Russell, *Dark Waters*, Livingston, MT: Clark City Press, 1988

Cholmondeley-Pennell, Harry, *The Modern Practical Angler*, Frederick Warne, 1870

Coates, Peter, *Salmon*, Reaktion Books, 2006

Coleby, R. J. W., *Regional Angling Literature*, Billinghay: printed for the compiler, 1979

Colquhoun, John, *The Moor and the Loch*, Edinburgh: William Blackwood, 1840

Couch, Jonathan, *A History of the Fishes of the British Islands*, 4 vols, George Bell & Sons, 1877

Cuvier, Georges (ed. Theodore W. Pietsch), *Historical Portrait of the Progress of Ichthyology, from its Origins to Our Own Time*, Baltimore: Johns Hopkins University Press, 1995

Dennys, John, *The Secrets of Angling*, W. Satchell & Co., 1883 reprint of 1613 original

Downes, Stephen, and Martin Knowelden, *The New Compleat Angler*, Orbis Publishing, 1983

Dunne, J. W., *Sunshine and the Dry Fly*, A. & C. Black, 1924

Ellacombe, Revd Henry Nicholson, *Shakespeare as an Angler*, Elliott Stock, 1883

Falkus, Hugh, *Sea Trout Fishing*, H. F. & G. Witherby, 1962

——, *The Sea Trout*, H. F. & G. Witherby, 1987

——, and Frederick Buller, *Falkus and Buller's Freshwater Fishing*, Macdonald and Jane's, 1975

Farson, Negley, *Going Fishing*, Country Life Ltd, 1942

Fedden, Romilly, *Golden Days: From the Fishing-log of a Painter in Brittany*, A. & C. Black, 1919

Fort, Tom, *Casting Shadows: Fish and Fishing in Britain*, William Collins, 2020

Francis, Francis, *A Book on Angling*, Longmans, Green & Co., 1867

Franck, Richard, *Northern Memoirs*, printed for the author, 1694

Gammon, Clive, *I Know a Good Place*, Shrewsbury: Swan Hill Press, 1990

Gierach, John, *Death, Taxes & Leaky Waders*, New York: Simon & Schuster, 2000

——, *All Fishermen Are Liars*, New York: Simon & Schuster, 2014

Gingrich, Arnold, *The Fishing in Print: A Guided Tour through Five Centuries of Angling Literature*, New York: Winchester Press, 1974

Goodspeed, Charles E., *Angling in America*, Boston, MA: Houghton Mifflin Co., 1939

Greene, Henry Plunket, *Where the Bright Waters Meet*, Philip Allan & Co., 1924

Grey, Sir Edward (Viscount of Fallodon), *Fly Fishing*, J. M. Dent, 1930, rev. edn of 1899 original

Grey, Zane, *Tales of the Angler's Eldorado: New Zealand*, New York: Harper & Bros, 1926

Hackle, Sparse Grey (Alfred W. Miller), *Fishless Days, Angling Nights*, New York: Crown Publishers Inc., 1971

Haig-Brown, Roderick, *A River Never Sleeps*, Collins, 1948

Hammond, Bryn, *Halcyon Days*, Shrewsbury: Swan Hill Press, 1992

Hare, C. E., *The Language of Field Sports*, Country Life Ltd, 1949

Hawker, Col. Peter, *The Diary of Colonel Peter Hawker*, 2 vols, Longmans & Co., 1893

Heilner, Van Campen, *Salt Water Fishing*, Philadelphia: Penn Publishing Company, 1937

Hemingway, Ernest, *The Old Man and the Sea*, New York: Charles Scribner's Sons, 1952

——, *The Nick Adams Stories*, New York: Simon & Schuster, Scribner Paperback Fiction edn, 1999

Hemingway On Fishing, ed. Nick Lyons, New York: Scribner Classics, 2000

Hemingway, Jack, *Misadventures of a Fly Fisherman*, Toronto: Key Porter, 1986

Herd, Andrew, *Angling Giants*, Ellesmere: Medlar Press, 2010

——, *The History of Fly Fishing*, Ellesmere: Medlar Press, 2011

Hersey, John, *Blues*, Weidenfeld & Nicolson, 1988

Higginbotham, James, *Piscinae: Artificial Fishponds in Roman Italy*, Chapel Hill: University of North Carolina Press, 1997

Hills, Rt Hon John Waller, *A Summer on the Test*, Philip Allan & Co., 1921

Hornell, James, *Fishing in Many Waters*, Cambridge: Cambridge University Press, 1950

Hughes, Ted, *Poetry in the Making*, Faber & Faber, 1967

——, *River: Poems by Ted Hughes, Photographs by Peter Keen*, Faber & Faber, 1983

——, *Collected Poems*, Faber & Faber, 2003

Jennings, Luke, *Blood Knots: A Memoir of Fishing and Friendship*, Atlantic Books, 2010

Johnson, Stephen, *Fishing from Afar*, Peter Davies, 1947

Kelson, G. M., *The Salmon Fly*, published by the author, 1895

Kingsmill Moore, T. C., *A Man May Fish*, Gerrards Cross: Colin Smythe, 1979, rev. edn of 1960 original

Kreh, Bernard 'Lefty', *Fly Fishing in Salt Water*, New York: Crown Publishers Inc., 1974

Lampman, Ben Hur, *A Leaf from French Eddy*, Portland, OR: Touchstone Press, 1965

Lang, Andrew, *Angling Sketches*, Longmans, Green & Co., 1891

Lang, Cecil ('Skene Dhu'), *The Angler in India*, Allahabad: Pioneer Press, 1923

Leeson, Ted, *The Habit of Rivers*, New York: Lyons & Burford, 1994

Luce, A. A., *Fishing and Thinking*, Hodder & Stoughton, 1959

Lyons, Nick, *Confessions of a Fly Fishing Addict*, New York: Fireside Books, 1989

McClane, A. J., *McLane's New Standard Fishing Encyclopaedia and International Angling Guide*, New York: Holt, Rinehart and Winston, 1965

McCully, C. B., *Fly-Fishing*, Manchester: Carcanet Press Ltd, 1992

McDonald, John (ed.), *The Complete Fly Fisherman: The Notes and Letters of Theodore Gordon*, Jonathan Cape, 1949

McGuane, Thomas, *Ninety-two in the Shade*, New York: Farrar Straus & Giroux, 1973

——, *The Longest Silence: A Life in Fishing*, New York: Alfred A. Knopf Inc., 1999

Maclean, Norman, *A River Runs Through It and Other Stories*, Chicago: University of Chicago Press, 1976

MacMahon, A. F. Magri, *Fishlore*, Penguin Books, 1946

Marinaro, Vincent, *A Modern Dry-fly Code*, New York: Crown Publishers Inc., 1950

Marinetti, Filippo Tommaso, *The Futurist Cookbook*, Penguin Classics, 2014 edn of 1932 original

Markham, Gervase, *The Pleasure of Princes, or Good Mens Recreations*, printed by John Browne, 1614

Marshall, Howard, *Reflections on a River*, H. F. & G. Witherby, 1967

Marson, Charles, *Super Flumina*, John Lane, 1905

Maxwell, Gavin, *Harpoon at a Venture*, Rupert Hart-Davis, 1952

Maxwell, Sir Herbert, *British Fresh-water Fishes*, Hutchinson & Co., 1904

Melville, Herman, *Moby-Dick; Or, The Whale*, New York: Harper & Bros, 1851

Mink, Nicolaas, *Salmon: A Global History*, Reaktion Books, 2013

Mottram, J. C., *Fly-Fishing: Some New Arts and Mysteries*, Field & Queen, 1915

Oppian, *Halieuticks of the Nature of Fishes and Fishing of the Ancients*, translated from the Greek by John Jones, Oxford: 'printed at the Theater', 1722

Owen, James, *Trout*, Reaktion Books, 2012

Parker, Eric (ed.), *An Angler's Garland*, Philip Allan, 1920

Patterson, Neil, *Chalkstream Chronicle*, Ludlow: Merlin Unwin Books, 1995

Paxman, Jeremy, *Fish, Fishing and the Meaning of Life*, Michael Joseph, 1994

Penn, Richard, *Maxims and Hints for an Angler and Miseries of Fishing*, John Murray, 1833

Picasso, Sydney, and Anthony J. P. Meyer, *Fish Hooks of the Pacific Islands*, Munich: Hirmer Publishers, 2012

Profumo, David, and Graham Swift (eds), *The Magic Wheel: An Anthology of Fishing in Literature*, Heinemann, 1985

Prosek, James, *The Complete Angler: A Connecticut Yankee Follows in the Footsteps of Walton*, New York: HarperCollins, 1999

Pryce-Tannatt, T. E., *How to Dress Salmon Flies*, A. & C. Black, 1914

Quarry, William W., *Salmon Fishing and the Story of the River Tweed*, Ellesmere: Medlar Press, 2015

Radcliffe, William, *Fishing from the Earliest Times*, John Murray, 1921

Raines, Howell, *Fly Fishing Through the Midlife Crisis*, New York: William Morrow & Co., 1993

Rangeley-Wilson, Charles, *Chalkstream*, Ellesmere: Medlar Press, 2009

Ransome, Arthur, *Rod and Line*, Jonathan Cape, 1929

——, *Mainly About Fishing*, A. & C. Black, 1959

Ritz, Charles, *A Fly Fisher's Life*, Max Reinhardt, 1972 edn

Roberts, Callum, *The Unnatural History of the Sea*, Washington DC: Island Press, 2007

Roberts, Morley, *A Humble Fisherman*, Grayson & Grayson, 1932

Santella, Chris, *The Tug Is the Drug*, Maryland: Stackpole Books, 2017

Schullery, Paul, *American Fly Fishing: A History*, New York: Crown Publishers Inc., 1950

——, *Royal Coachman: The Lore and Legends of Fly-Fishing*, New York: Simon & Schuster, 1999

Schwabe, Calvin W., *Unmentionable Cuisine*, Charlottesville: University of Virginia Press, 1979

Schwiebert, Ernest, *Remembrances of Rivers Past*, New York: Macmillan Company, 1972

Scrope, William, *Days and Nights of Salmon Fishing in the Tweed*, John Murray, 1843

Senior, William, *Lines in Pleasant Places: Being the aftermath of an old angler*, Simpkin, Marshall & Co., 1920

Sheringham, Hugh Tempest, *An Angler's Hours*, Macmillan, 1905

——, *Coarse Fishing*, A. & C. Black, 1913

Skues, G. E. M., *The Way of a Trout with a Fly*, A. & C. Black, 1921

Sparrow, Walter Shaw, *Angling in British Art*, John Lane, 1923

Spencer, Sidney, *Newly from the Sea: Fishing for Salmon and Seatrout*, H. F. & G. Witherby, 1934

Thomas, Dylan, *Collected Poems 1934–1952*, J. M. Dent & Sons, 1952

Thoreau, Henry David, *Walden, or Life in the Woods*, Boston, MA: Ticknor and Fields, 1854

Traver, Robert, *Trout Madness*, New York: St Martin's Press, 1960

Van Dyke, Henry, *Fisherman's Luck*, New York: Charles Scribner's Sons, 1899

Venables, Bernard, *Mr. Crabtree Goes Fishing*, Daily Mirror, 1949

Walton, Izaak, and Charles Cotton, *The Compleat Angler or the Contemplative Man's Recreation*, published by Richard Marriot, rev. 5th edn, 1676

Westwood, Thomas, and Thomas Satchell, *Bibliotheca Piscatoria*, W. Satchwell, 1883. With Supplement by R. B. Marston, Sampson, Low, Marston & Co., 1901

White, T. H., *England Have My Bones*, Collins, 1936

Wigan, Michael, *The Salmon*, William Collins, 2013

Wiggin, Maurice, *The Passionate Angler*, Sylvan Press, 1949

Williams, Alfred Courtney, *Angling Diversions: Angling History and Humour*, Herbert Jenkins, 1945

Williamson, Henry, *Salar the Salmon*, Faber & Faber, 1936

——, *A Clear Water Stream*, Faber & Faber, 1958

Wormald, Mark, Ned Roberts and Terry Gifford (eds), *Ted Hughes: From Cambridge to Collected*, Basingstoke: Palgrave Macmillan, 2013

Wulff, Lee, *The Atlantic Salmon*, New York: A. S. Barnes & Co., 1958

Yates, Christopher, *Casting at the Sun*, Pelham Books, 1986

Yeats, W. B., *Collected Poems*, Macmillan, 1933

Znamierowska-Prüffer, Maria, *Thrusting Implements for Fishing in Poland and Neighbouring Countries*, Warsaw: Scientific Publications Foreign Cooperation Center of the Central Institute for Scientific, Technical and Economic Information, 1966

Zern, Ed, *Hunting and Fishing from A to Zern*, New York: Nick Lyons Books, 1985

ACKNOWLEDGEMENTS

In writing a book of this nature, which has gradually taken shape over several years, I have had assistance, inspiration and input from so many people that I cannot possibly thank them all individually by name. However, in these few final lists, I would like to express my heartfelt appreciation, starting with those whose companionship has meant a great deal, as has their sharing of stories and experiences on the water: Evan Bazzard, C.D. Clarke, Bill Forse, Tom Fort, Brian Fratel, Maureen Gilmour, Fergus Granville, Loyd Grossman, Janie Grosvenor, Max Hastings, Veronica Hotchkiss, Einar Falur Ingólfsson, Charles Jardine, Jonathan Lawrence, Simon McKay, Robert Montague, William Neville, Neil Patterson, Jeremy Paxman, Trevor Potter and Dana Westring, Charles Rangeley-Wilson, John Rocha, Michael Samuel, Chris Sandford, Glyn Satterley, Victoria Sheffield, Tom Stoppard, Richard Viets, Todd Warnock and Sebastian Whitestone.

On the home front, there have been many happy times with my brother Mark, cousins Martin and Mary Ann Janson, all my Fraser in-laws and Hugh Clement. A special mention in dispatches must go to Alasdair Fraser ('The Doctor'), who has been an impeccable sporting friend for more than four decades, and has stoically endured most of my tall tales. I have loved fishing with James, Tom and Laura, and long for more days to come. I must also pay tribute to my late mentors: Harold Balfour, George Murray, Megan Boyd, Dermot Wilson, Hugh Falkus, Arthur Oglesby and Fred Buller.

For their generosity in inviting me to fish (often in their company) or for arranging expeditions, and many other kindnesses, I am grateful to: James Abercorn, James Adeane, Jamie Barshall, Peter Baxendale, Archie Boyd, Alastair and Sheila Brooks, Jonathan Bulmer, Adrian Dangar, William Daniel, Bill and Pam Doyle, Mark Firth, Neil Freeman, Richard Gledson, Andrew Graham-Stewart, David and Linda Heathcoat-Amory, the late David Hodgkiss, Bo Ivanovic, Miles Larby, Anthony Lavers, Niall Leveson-Gower, Abe Lyle, Justin McCarthy, Ross Macdonald, Wayne McGee, Mat McHugh, Peter McLeod, Peter Mantle, Stephen May, Tarquin Millington-Drake, Mike Mirecki, Innes Morrison, Henry Mountain, Ian Neale, Mark Newton, Ralph and Jane Northumberland, 'Bumble' Ogilvy-Wedderburn, Pepe Omegna, Harald Oyen, Algernon Percy, Tim Pilcher, Robert Rattray, the late Guy Roxburghe, Vladimir Rybalchenko, David Stewart Howitt, Justin Maxwell Stuart, Jamie and Sarah Troughton, the late Orri Vigfússon, Andrew Wallace, Toby Ward, Charles and Antonia Wellington, Anita Wigan, Michael Wigan, and Óttar Yngvson.

On the water, I have spent many formative hours and have learned a great deal from the expertise and guidance of numerous experts and professionals, many of whom have become friends in the process. This list is incomplete, but includes: Gilly Bate, Teihotu Brando, John Buskie, Sean Clarke, Arthur Deane Sr, Rod Dixon, Robin Elwes, Colin Espie, Peter Fraser, Ingi Helgason, Peter Hellard, George Inglis, Robert King, Ronnie McElrath, James MacLetchie, Phil Parker, Pedro Maria Peleaz, Oliver Rampley, Tony Scherr, Reuben Sweeting, Howard Taylor and John Taylor.

For helping me steer a literary course between the Rocks of the Bosphorus, huge thanks for encouragement and advice in various crucial ways from: Keith Elliott, Luke Jennings, Tom McGuane, Graham Swift and Mark Wormald. The staff of both the London Library and the British Library have once again given invaluable assistance with my research. An assortment of Editors – past and present – have also sanctioned or commissioned work that has allowed me to travel to so many exciting places: Richard Addis, Jim Babb, Mark Bowler, Simon Courtauld, Andrew Flitcroft, Patrick Galbraith, Geordie Grieg, Mark Hedges, the late Peter Lapsley, Paula Lester, Sandy Leventon, Victoria Mather, Sandy Mitchell, Sarah Spankie, Rupert Uloth, Merlin Unwin, and Chris Yates. I owe practically a lifetime's debt to Michael Meredith, John Fuller, Bernard O'Donoghue and the late Emrys Jones for trying to teach me some of the finer points about how to write, and read.

For their generosity with copyright permissions, I want to thank Carol Hughes, Tom McGuane, Colin Smythe, Anne Williamson and the Estate of Norman Maclean.

The seed crystal for this book dates back to the time when the late Gillon Aitken was my agent, though subsequently the project was ably brought to fruition by my friend (and erstwhile editor) Peter Straus, who has my sincerest thanks – as do Matt Turner, Eliza Plowden, and other colleagues at the Rogers, Coleridge & White Agency. Chris White, who commissioned the manuscript in his early days at Scribner, proved to be an astute and exemplary editor, urging me towards the finishing line with every sign of commitment, aplomb, enthusiasm and good humour (and virtually no Procrustean inclinations). I am profoundly grateful to him, and also Kaiya Shang, Francesca Sironi and Art Director Matt Johnson. Rhiannon Carroll and Amy Fulwood did sterling work with PR and Marketing. Charlotte Chapman meticulously copy-edited the text, and set me straight on several slips and ambiguities: any remaining mistakes are mine.

It is traditional to reserve the final glorious mention for your partner in life – but really, my wife Helen deserves pride of place. For her patience and support during seemingly endless procrastination, travails and temperamental ditherings down the decades, I can only offer her my enduring devotion and thanks.

DP,

Perthshire, November 2020.

Permissions

The author and publishers have made every attempt to contact copyright holders, but if they have inadvertently overlooked any they will be pleased to make the necessary arrangements in any future editions. They gratefully acknowledge permission from the following sources to reprint material in their control:

Courtesy of the Ted Hughes Estate, permission to quote two unpublished prose quotations, along with extracts from *Poetry in the Making* (1967) published by Faber and Faber; from *Collected Poems* (1983) ed. Paul Keegan, also published by Faber; likewise from *Flowers & Insects (1986), A Rain-Charm for the Duchy* (1992), and *Letters of Ted Hughes* (2007) ed. Christopher Reid.

Extract from *The Longest Silence* by Thomas McGuane (UK edition 2000) published by Yellow Jersey Press, copyright Thomas McGuane 1999, by kind permission of the author.

Extract from *A Clear Water Stream* by Henry Williamson (1958), published by Faber and Faber, courtesy of the Henry Williamson Literary Estate.

Extract from *Collected Poems* by Dylan Thomas (1952) published by JM Dent & Sons, copyright The Dylan Thomas Trust, by kind permission of David Higham Associates.

Extract from *A Man May Fish* by Kingsmill Moore (1979 edition) published by Colin Smythe Ltd. By kind permission of Colin Smythe Ltd.

Extracts from *A River Runs Through It and Other Stories* by Norman Maclean (1976) published by the University of Chicago Press © 1976 by The University of Chicago. By kind permission of the Norman Maclean estate.